THE TEMPEST WITHIN

AN ACCOUNT OF EAST PAKISTAN

BY THE SAME AUTHOR

POETRY
A Beginning
Poems
John Nobody
The Brass Serpent
A Collection of Poems, 1955-1970 (in press)

PROSE
Gone Away
My Son's Father
From East and West (in press)
The Seventh Person (in press)

Dom Moraes

The Tempest Within

An Account of East Pakistan

BARNES & NOBLE, Inc.
NEW YORK
PUBLISHERS & BOOKSELLERS
SINCE 1873

First Published in the United States, 1971
by Barnes & Noble, Inc.
ISBN 389 04201 3

PRINTED IN THE UNITED STATES OF AMERICA

FOR

DR AND MRS RAMAIAH NAIDU

with gratitude and affection

PREFACE

This book is an account of a short visit I paid to East Pakistan soon after the cyclone that devastated its coastal areas in November 1970. I was on an assignment for the New York Times Magazine and my brief was purely to cover the aftermath of the cyclone. My trip, however, overlapped with the elections in early December, and there was political excitement in the air. It was a dual cyclone, I felt; a whirlwind of politics accompanying the disaster that came from the sea. It was the first time I had visited the country, and I knew little about the political situation, but it interested me deeply, and I started to read up on it, and talk to people about it. This volume is the result. I do not pose as an expert on Pakistan: if this book has any value, it is as a personal account of what one writer saw and felt during a necessarily short stay in an area on which very little firsthand information is available in India.

For the section of the book where I discuss the political development of East Pakistan, I am indebted to several authors whose volumes I used for research. These volumes are : *India, Pakistan and the West*, by Percival Spear (Oxford University Press, 1949); *The Making of Pakistan*, by Richard Symonds (Faber, 1950); *Asia and Western Dominance*, by K. M. Panikkar (Allen & Unwin, 1953); *The Discovery of India*, by Jawaharlal Nehru (Meridian Books edition, 1956); *Glimpses of Bengal in the Nineteenth Century*, by R. C. Mazumdar (privately printed, 1960); *East Pakistan*, by D. N. Banerjee (Vikas Publications, 1969); and *Pakistan in Crisis*, by G. S. Bhargava (Vikas Publications, 1969).

I wish to thank the New York Times Publishing Co. for allowing me to reproduce in this text, in a somewhat adapted form, sections of an article on East Pakistan which appeared in the New York Times Magazine of 10 January 1971.

DOM MORAES

Contents

Chapter I

PLANE TO PAKISTAN

Getting to East Pakistan presented no problem. My pockets were stuffed with expenses, I needed no papers or visa, and a Thai Airways plane departs from Delhi thrice weekly for Dacca. I boarded it on an afternoon of wind and dust, and as the monuments and ministries of the Indian capital dwindled under us, resigned myself to the routine of air travel: the pages of the paperback flicked over one by one, the timed arrival of food on a plastic tray, the occasional admonitory quack of the loudspeaker. The routine sealed us into a private world: when, an hour out of Delhi, the sleepy plane droned like a harvest bee past the plumed helmet of Mount Everest, nobody even bothered to look out. After a while, the loudspeaker addressed us afresh, to say that we had crossed the East Pakistan border. The country below us, wracked by cyclones both natural and political, was to be the arena of our work, for most of the passengers were foreign correspondents. This time we all looked out, anxious to see what East Pakistan was like, even from the air. It looked gloomy and bereft. A lurid and streaky sunset had spilled its yolk over a torpid brown landscape seamed and patched with yellow waterways and creeks. Far off ahead, across the darkened plain, the lights of Dacca trembled slowly towards us, like a constellation of fallen and scattered stars.

East Pakistan squats back in swampland off the Bay of Bengal. Its 55,000 square miles are mostly flat, a cobweb of canals and slow, deep rivers which empty into the sea. Jute and rice flourish in the damp black soil. It is a land of waterways and villages, with little industrial or urban life. Dacca, the capital, is also the only real city.

The plane roared on the runway, then came silence and the stop. Outside, the air was moist and warm, and the first prickly stars had come out. Hundreds of people cluttered the airport steps, mute, whiteclad, watching us walk toward them. I asked a Pakistani official who they were. "They have come to watch the planes," he said. "They have no other occupation, you see. There are very many unemployed in Dacca." It was not a new fact. Unemployment and poverty have been traditional, almost, in East Bengal for the last 150 years. Before that it was a prosperous province with full employment. The east was a ricebowl, and the weavers of Dacca, hunched like dark spiders over warp and weft, spun cotton into the finest muslin in the world. It was exported to ancient Rome and China—the Renaissance knew of it, and, muslin apart, quantities of spices and other important articles of trade floated down the waterways of East Bengal to the rich ports which abutted on the stormy but sail-filled sea.

The attraction of wealth, however, is universal : especially when that wealth belongs to a frail and unwarlike people. In the 13th century, the Turki hordes swirled down into Bengal and occupied it. For more than 500 years thereafter, the province remained under Muslim rule. Naturally enough, it changed in some ways, notably in the religious beliefs of the people. The caste system introduced by the Aryans was by this time 3,000 years old, and the centuries had fossilized it. The untouchables and those of low caste, trapped on this unwieldy treadmill, now thought Islam, preaching the equality and brotherhood of its adherents, the only means of escape from an eternally inferior status in society. Mass conversions took place. These were often group conversions, where an entire village or professional community turned Muslim. Thus certain professions became largely the province of Muslims: weavers, tailors, and butchers, for example, all followed Islam. The Muslim state protected them from the retaliatory wrath of the Brahmins.

The highly born Hindus, however, were in the main remarkably unaffected by the influence of Islam. As the historian R.C. Mazumdar wrote dolefully, "In particular it is strange that the wonderful social democracy of the Muslims, far from removing the barriers artificially planted between man and man, rather made them stronger and stronger" : at least as far as the caste Hindus were concerned. Religion divided the province of Bengal, with the eastern population mainly Muslim, and the western population mainly Hindu.

It is a curious fact that though, wherever the Muslims were conquerors, they arrived as austere desert warriors, carrying the instruments of purification, fire and the sword, their descendants suffered a rapid decline, turning into cruel, slothful, lecherous, perfumed sybarites in a matter of decades. For the Muslims, as Jacques Barzun has said of the Americans, it was only a step from barbarism to decadence. The East Bengali nobles had already taken this step when in the 17th century the East India Company first established its posts in their province. "Bengal . . . lies open to the merchant and the soldier," writes Dr Percival Spear, "but its climate, its rivers and rains have combined to give it an atmosphere which envelops each conqueror or adventurer in turn, and a temperament which is sooner or later caught by all who stay there long." Whether one blames it on the weather or the Muslim character, the fact remains that by the time the Company came, the native nobles of the province were a far remove from their puritanical and hardy ancestors. They had expensive tastes, and most of them were therefore chronically broke.

There was money to be had, however, from the Hindu merchant princes who monopolized the river trade. They loaned it to the Muslims, either at very high interest or in return for favours. Their rupees propped the rickety thrones of successive rulers. The Company, meanwhile, waxed fat on overseas trade, and, to maintain its position, started cautiously to play off the political cliques of the Bengal court against each other. In 1756 the current ruler, Siraj-ud-Daulah, who was incensed by British support of his opponents, captured the Company fortress at Calcutta, but he didn't hold it for very long. Unwisely, as it proved, he quarrelled with some of his Hindu financiers. They proceeded to parley with the British, and to buy out the nobles and military officers around Siraj-ud-Daulah.

The Battle of Plassey, which took place on 23 June 1757,

was the result. "It was a transaction," writes K. M. Panikkar, "not a battle." The generals of the Muslim army, heavily bribed, refused to fight, and the Company troops under Robert Clive routed a shaken and decimated enemy. One of the defectors from the Muslim side, Mir Jafar, was appointed the Nawab of Bengal. He was the glove in which the hand of the British moved. He signed a document whereby the Company became, in effect, the landlord of the rich 24 Parganas district near Calcutta. The Mogul Emperor in Delhi became rather alarmed by these developments, but he was typically slow to act. Eventually, in 1764, he sent troops against the British, but his forces were defeated at Buxar. He was compelled to cede the revenue administration of three wealthy provinces, Bengal, Bihar, and Orissa, to the Company. The administrative policy of Clive towards these provinces was exceedingly simple: its aim was to milk the country for all it was worth.

The Bengal officials of the Company did so with a will. They forcibly took food-stuffs and other commodities from their producers at a price well below the normal market rate. They then forcibly sold them to the helpless inhabitants at a price far above it. The muslin weavers of Dacca suffered particularly, and legend has it that some of them cut off their right hands rather than work for the Company.

The peasants suffered most. Before the Company came to power, an Indian village operated on a cooperative basis, like an Israeli *kibbutz*. The peasants tilled fields which were, in effect, common property. Out of the harvest, a tax was paid to the ruler, and the remainder shared out among the villagers. The British, however, brought in a new system. A Collector, or landlord, was appointed to administer large tracts of property. He was taxed by the British on the revenue he collected. It was thus in the interests of both the British and the new landlords, who were mostly Hindus, to collect this revenue extensively and quickly. The landlord pinioned the peasants under a plump, iron heel, and they entered into a kind of serfdom, living on rented land, the yield of their crops snapped up by the authority above.

In 1769, only five years after Buxar, Richard Becher, a Company official, wrote to the head office in London: "It must give pain to an Englishman to think that since the accession of the Company the condition of the people of this country has been worse than it was before. This fine country, which flourished under the most despotic and arbitrary government, is verging towards ruin." A year after he wrote this, three million starved to death in the Bengal famine of 1770. The landlords of East Bengal, in a petition to London, lamented: "There is now scarce anything left in the country." By that time, they were perfectly right. Wealth was pouring out of India, and it must have been scant comfort to those from whom it was taken to reflect that it was serving a very useful purpose in England, where it financed the Industrial Revolution.

The indigo plantations which now started in Bengal and Bihar were another evil. The planters were English, but says Edward Thompson, "They were a lawless set, who combined some of the worst features of the 18th century *zamindars* with the tenacity of the Indian moneylender. Very few grew indigo themselves. They obtained their raw materials by advancing money to cultivators and gradually getting them completely under control." The Official Indigo Commission's report stated: "It matters little whether the ryot took his original advance with reluctance or cheerfulness. The result in either case is the same. He is never afterwards a free man." Something like the slave system of Alabama developed in the lush fields of eastern India.

The British in India normally favoured the Muslim population, who were thought to be more honest and trustworthy than the Hindus. The exception to the rule was in Bengal. The Hindus here were intelligent, volatile, and often rich. The Muslims, by and large, were'nt. The peasants were serf-like, under Hindu masters: they could not better or educate themselves because they were so poor. Even the lot of the richer Muslims, under the Company, was a hard one. Sir William Hunter, a Bengal civilian of the early 19th century, wrote of "Lower Bengal...in which...the Muhammadans have suffered most severely under British rule.... If ever a people stood in need of a career, it is the Musalman aristocracy of Lower Bengal...." Hunter points out that these people had once administered the taxation, officered the police, monopolised the judiciary, and drawn pay from the State for the peasants whom they forced into the Army. These perquisites had now vanished, and, says Sir William, "170 years ago, it was almost impossible for a wellborn Musalman in Bengal to become poor, at present it is almost impossible for him to continue rich."

The Company thus had no profit to derive from the Muslims, and some from the Hindus. The capital of Bengal had always been in the Muslim east, mainly at Dacca: the Company moved it to the Hindu west, to Calcutta. The Hindus, as always, had a finger in the pie of commerce, and since many of them were educated, they had a virtual monopoly of clerical posts under the Company. The Muslims meanwhile sweated it out in the east, in the ricefields of the Hindus or the indigo plantations of the British. When the Manchester textile mills became interested in the Indian market, the British forced the muslin weavers to stop production. By 1854 the looms of Dacca were mostly silent, and hundreds of people had been thrown out of employment.

It was therefore hardly surprising that the Muslims were both spiritually and physically reduced by their privations, but the British seldom seem to have analysed the causes of their decline. Charles Grant, in a book written in 1792, described the peasantry of East Bengal as inferior to the most backward classes in Europe. Lord Hastings, in 1813, said they were "limited to mere animal functions, and even in them indifferent. . . . [They have] the dexterity which any animal of similar conformation but with no higher intellect than a dog, an elephant or a monkey might be supposed capable of attaining." The Hindu peasantry of the west, it must be added, was little better off and sunk in depression and despair, the mass of Bengalis remained mutely indifferent to the succession of political upheavals which brought about the Sepoy Mutiny of 1857.

The shock of the Mutiny, the notorious crookedness of Company officials, and the undoubted profits available to anyone who ruled India, all played a part in the Crown decision of 1858, whereby the British Government took over the country. They ruled fairly, if strictly, at least in comparison with the Company, but some of the abuses of power perpetrated by their predecessors now came to roost. In 1859, the indigo workers at last rebelled. They refused to sow the crop, and riots followed in Bengal. The Blue Mutiny, as the indigo revolt was called, had more effect in the province than its namesake. It kindled the first raw sparks of political aspiration in the Muslim masses, an aspiration which nearly a century later was to lead them to the further tragedy that is East Pakistan.

The airport was a noisy, excited mill of people. The cause of excitement was not immediately apparent. Nothing was happening, except that the passports of the passengers from Delhi were being scrutinised and stamped, but all around us were yelling, waving men. They were mostly short, with broad, high cheekboned faces, caprine beards, and skullcaps. Gradually I began to understand their motives: they were touts, offering us women, liquor, and an illegal rate for pounds and dollars.

The official who stamped my passport was affable and young. He smiled at me across the counter. "You are British?" he said. "You are not Indian? Good. Indians are rubbish." He then added, "You know, you are press person, not tourist, you will only get four rupees per dollar at the bank. But I have a good friend who will give you tourist rate. Eight rupees per dollar. I will finish my duty, then I will show you." Long practice has made me immune to such offers. I shook my head.

Another crowd swarmed around the exit. The touts here had taxis to offer. I selected one, and, accompained by a colleague from Reuters, climbed into it and demanded the Intercontinental. The driver stamped on the accelerator as though it was a caste Hindu, and we roared off into the darkness. The pavement foodshops of the city were open. A garlic smell rose off the cauldrons of meat, lentils, and rice that stewed over charcoal fires. The crowded streets were filled with the wail of radio music. The cries of the vocalists had a prolonged and strident quality as intensely Islamic as the onion domes and phallic spires which flashed past in silhouette on the luminous sky. Dacca possesses more mosques than any other city on the subcontient. The darkness made them dramatic: during the day, as I discovered later, they have a tatty and decrepit look. The fact that we were in a Muslim country was further underlined by the absence of women from the crowded pavements. The few who were around, clad in burkhas, resembled small black tents.

Beyond a huddle of tenements, the huge white bulk of the Intercontinental lifted its burden of bedrooms, restaurants, bars, and unlimited expenses accounts into the dark sky. We climbed out of the cab. The driver demanded a dollar. We offered him the equivalent in his own currency, but he shook his hand and leered wisely through his beard. "No," he said. "I want American dollar. Dollar good money." We gave him the dollar and followed our suitcases into a vast and crowded lobby. It resembled the departure lounge at an international airport, and was full of slightly dishevelled, elaborately casual men with cigarettes clamped intensely between their lips. Their nationalities and languages were colourfully varied, but we recognised our own kind. "Christ!" said the Reuters man, "this place stinks of foreign correspondents!"

We elbowed our way to the reception desk, where a harassed clerk confirmed our reservations. A hot bath and drink represented the limits of my immediate ambitions. I went upstairs and had a bath, then came back down to have the drink. The red carpet that led to the bar was well-trodden, like the path to a forest waterhole. The press, after all, had been in the hotel two weeks already.

In 1860, the year after the Blue Mutiny, Nawab Abdul Latif, a Qazi from the Faridpur district of East Bengal, started the Mohammedan Literary Society in Calcutta. Though this hardly sounded a revolutionary move, it had widespread effects. The Society was active in the cause of social and educational uplift for the Muslims. Its efforts helped to create a new Islamic elite in Bengal. More directly political moves came from the Central National Mohammedan Association, founded by Syed Amir Ali in 1877. This organization, in 1882, presented the then Viceroy, Lord Ripon, with a detailed list of Muslim complaints for which it sought redress. By this time the spread of education in Bengal, though brought about by the British, had become rather inimical to their interests. An educated class, mostly Hindu, had emerged. It was concentrated in Calcutta, and here its members clamoured for the end of British rule. The concurrent emergence of a Muslim elite disquieted the British. It was clearly necessary to prevent the two communities from allying with each other. The Muslim complaints of Hindu domination provided Lord Curzon, Viceroy in 1905, with the opportunity for a master stroke of policy. He delighted the Muslims, infuriated the Hindus, and served the interests of the British, by dividing Bengal into two. The Muslim East was ruled from the old capital of Dacca, the Hindu West from Calcutta.

The year after partition, 1906, was historically even more important. In this year Nawab Vaqar-ul-Mulk, himself an East Bengali, convened a meeting of Muslims from all over India to discuss education. The discussions on education turned out to be the least important part of the conference, which took place in Dacca: for on 30 December 1906, the Nawab of Dacca moved a resolution for the foundation of a Muslim political organisation. This organisation was to be known as the All-India Muslim League, was to protect and advance the political rights and interests of Indian Muslims, and pledged loyalty to the British Government. The concept of the Muslim League came from East Bengal. "Several hundred gentlemen," wrote Mohammed Noman, a contemporary commentator, "assembled at Dacca. . . thousands came from the province of East Bengal itself." A Drafting Committee, to frame a Constitution for the League, was appointed, and included three East Bengalis.

But, anticipating already the pattern to be enacted in Pakistan, the East Bengalis were soon to suffer what they considered to be a betrayal by the Muslims of the west. The League, founded in Dacca though it was, held its first actual session in Karachi in 1907, and its second in Aligarh in 1908. The pattern held, in that its meetings thereafter were almost always in the United Provinces, where the Muslims spoke Urdu and were ethnically different from those of East Bengal. (Nehru, indeed, estimated that 98 per cent of Bengali Muslims were the descendants of Hindu converts.) It was felt in Dacca that the Muslims of the west considered their own problems to be of paramount importance, and attached little importance to the special and different difficulties of their brethren in Bengal. This idea was substantiated in 1911. In this year the British, submitting to heavy pressure from Hindu power blocs in Calcutta, annulled the partition of Bengal. The Muslims were appalled at this, and there were violent protests. But the League, at this critical moment, maintained silence. The Muslims of the west made no move of sympathy or solidarity towards the east.

Sympathy, indeed, came from Britain itself. In February 1912, Lord Minto lamented in the House of Lords: "We told the Muslims that partition was a settled fact. We assured the Muslim population of West Bengal of our appreciation of their loyalty and our determination to safeguard their interests." The loyalist leaders now started to try to milk concessions out of the British, as a kind of compensation for their betrayal. They obtained concessions in the field of education. It was further established that in the field of public appointments, a third of the posts available should be allotted to qualified Muslims, even if more highly qualified Hindus were to apply for them. The most important concession, however, was in the area of property ownership. In 1880 the British had the impression that the Hindu zamindars, tied to them by vested interests, would be loyal.

Of what happened 30 years later, Dr Spear writes: "The Government attitude to the Muslim community was now very different to what it had been after the Mutiny. As a result, six special constituencies of landlords were created for the Imperial Legislative Council....This measure, which seemed but an expedient to secure the representation for an important body of opinion, was one of the deepest import. It may perhaps be described as the official germ of Pakistan."

The new Muslim landlords were often the possessors of British titles, and they included several loyalist political leaders. The younger politicians, however, were stirred to wrath by the abolition of partition. Fazlul Huq was one of their chief spokesmen and he represented the educated elite, a quite different class from the landed aristocrats and titled nobility who led the loyalists. His group concurred with the Muslim League when, in its 1913 constitution, its goal was declared to be the attainment of a system of self-Government suitable to India. Mohammad Ali Jinnah became a League member in this year, declaring that "loyalty to the Muslim League and the Muslim interest" would not overshadow "the larger national cause to which his life was dedicated."

The outbreak of the Great War in 1914 caused a new shift in the Muslim stance. Britain was fighting Turkey, amongst others, and the Khilafat was centred in Turkey. This alienated many Indian Muslims from the British, and it caused some to draw close to the Indian National Congress, which hitherto had been dominated by Hindus. In 1916, for example, the League and the Congress shared a political platform. Fazlul Haq presided over the All-India League session, and was also General Secretary of the Congress. The Pan-Islamic Khilafat movement, however, dominated the Muslims during the war years. The Khilafat looked on Islam as a world brotherhood, whose members owed no special allegiance to any country. The loyalists looked on Islam as a political identity: they wanted to protect the interests of the Muslim minority through the British in India. Between these blocs came the educated elite, led by Jinnah, who looked on Indian nationalism as a means of bringing about political and economic equality for Muslims. It was a case of pull devil, pull baker: the three contradictory wings of opinion created disarray in the League.

Meanwhile, in Bengal, Fazlul Huq created the Krishak Proja Party, a purely regional affair, which throve on the social and economic compulsions that drove the Bengali Muslim. After the collapse of the Khilafat movement in 1924, Fazlul Huq's party increased its power. Jinnah, simultaneously, saw the Khilafat collapse as an opportunity to reinstate the Muslim League as the only spokesman of Indian Muslims. He convened a meeting in Lahore in 1924. But the Muslims were still divided, and the open session in March 1928 ended in chaos. The nationalists and loyalists broke away into separate parties, and Jinnah, like the Scholar Gypsy, fled their touch, their feverish contact fled, and went to London to set up in practice as a barrister.

The hotel bar was, for some reason, in utter darkness. This did not stop the customers: they perched on stools around the long sweep of the bar, or sat at tables, served by indistinct waiters, and talked. Presently we had a little light on the scene: a series of bulbous lamps opened their eyes in the dark, like owls, and revealed us to each other. The drinkers were mostly correspondents, some of whom I knew and hailed. We gathered at a providentially empty table, and talked of the elections that were about to take place. Some of the others had been in Dacca for days, covering the cyclone which, on 12 November 1970, had ripped into East Pakistan: the worst natural disaster of the century, drowning and killing hundreds of thousands of people, destroying millions of rupees worth of livestock and property. They were hard men, professionals, but the magnitude of the disaster had shaken even them. They shook their heads over it, and had another drink.

After a while I wandered out into the lobby to buy a book, since I have the expensive vice of being unable to sleep unless I read. As I examined the wares of the bookstall, which seemed to offer nothing less abrasive to the aesthetic sense than Barbara Cartland, a young Pakistani came up behind me and tapped me on the shoulder. "You are Mr Moraes," he said. "I have read your books, and I recognize you from your pictures. Let me introduce myself. I am Rahman. I am a Government servant. I had some work here tonight, and I was about to leave when I recognised you." He was short, like most Bengalis, with a wide friendly face and thick, rather coarse black hair, heavily oiled, and was very young. "Please," he said impulsively, "my friends are poets, they would be overjoyed to meet you, let me take you to them." I hesitated. Another drink, dinner, and bed issued an imperative summons to the hedonist in me, but I knew I would offend Rahman if I refused, and also this could be of interest. So I said I'd come.

He led me, chattering delightedly as he did so, through the swing doors into the warm garlicky dust of the night. "What fortune! What fortune, to find you here! I also have tried to write poetry, but it is very poor. I have enough intelligence to know that. So I console myself with the company of poets. *All* my friends are poets. *All*, all. The friend to whom I will take you, he is editor of one poetry magazine. Come, we will take a cycle rickshaw. I cannot afford a taxi, and tonight you are my guest." A dark, scrawny man pedalled a rickshaw up to us. We clambered in, and huddled together on the narrow, luridly vermilion seat. The canvas hood above us was adorned with colourful tassels which flapped into my eyes. The driver, groaning and lunging over the pedals as though in some painful sexual paroxysm, took us unsteadily out into the wide boulevard that fronted the hotel. "Hah!" said Rahman. "See that fellow! How thin! hah, how miserable! Not like the tall handsome people of West Pakistan. That is what these West Pakistanis tell us." He pointed out ridge upon ridge of shanties to our left, made of canvas sheets propped on poles, and smiled. "Right in front of the Intercontinental! The East Pakistanis live in such places as these, but the West Pakistanis live in the hotel. They are eating fine food. These wretched people, like him," nodding at the rickshaw man, panting over the pedals, "they are so low that all they have is a handful of rice a day. Yes, Mr Moraes, we are terrible people, we East Pakistanis, like animals, and we deserve our fate."

He laughed bitterly, waving his hand at a spired dome off the road. "We are very rich, we East Pakistanis. Rich in mosques only. Also, of course, in cyclones." A nervous tic seemed to have taken hold of him, he could talk of nothing except the miseries of his province. His manner was slightly hysterical, and his irony clumsy. He was trying to explain something he felt violently about, and yet trying to conceal what he felt. There was a sudden sad look about him.

Meanwhile we turned off the wide silent avenues, into narrow alleys. The tumbledown houses abutted on dirty pavements, kerosene lamps flared in the shop fronts, still open despite the hour, and people flooded around us. They were nearly all men, and as Rahman had said they were a different physical type from the tall, fair-skinned West

Pakistanis. Their faces behind their beards were Mongol and flat: they wore checked sarongs. East Pakistan abuts on Burma and Assam after all, and these people were visibly natives of South East Asia, as the Pakistanis of the west are visibly related to the Middle East. We drew up in front of a tall tenement, the ground floor of which was occupied by foodshops. The rancid smell of mutton fat and biryani rose from them. Over the fetid steam of the cauldrons, the saronged cooks meditatively picked their noses. "These are very good foods here," Rahman cried enthusiastically. "You must eat your dinner with us." His dark mood was forgotten: he was radiant as a child. He swept me into the tenement, up a noisy and narrow flight of wooden stairs. "What a wonderful evening this will be!" he said.

The concept of a separate Muslim state was first enunciated by a poet. Sir Mohammed Iqbal was the rather unlikely President of the Muslim League in 1930. On 29 December that year, he said: "I would like to see the Punjab, NWFP, Sind, and Baluchistan...a single state. Self government, within the British Empire or without the British Empire, the formation of a consolidated North West Indian Muslim State, appears to me...the final destiny of the Muslims at least of North West India." Iqbal clearly saw this Muslim state as part of India, but he made no suggestion as to the fate of the Muslims of Bengal.

The first Round Table Conference was held in London in 1930. It was followed in quick succession by two others. The Congress didn't attend the first conference, though Gandhi represented it at the second one. The Muslim League, however, was heavily represented at all three conferences. Though neither the Congress nor the League obtained much satisfaction from the Round Table discussions, a concept that made history was propounded during them by Chaudhuri Rahmat Ali, a Cambridge undergraduate, one of the numerous Muslim students who took part. He proposed the idea of Pakistan, a state which would free Islam from the Indian yoke. "The clash," he said, "is neither inter-religious nor intercommunal, nor even economic. It is, in fact, an international conflict of two nations' ambitions—Muslims for survival, and Hindus for supremacy.". Rahmat Ali, though opposed to Iqbal's concept of a Muslim state within India, agreed with him as to the composition of this state. He included within it, like Iqbal, the Punjab, NWFP, Sind, and Baluchistan, but, prophetically, he also included Kashmir. Eventually, in 1937, when Fazlul Huq's Krishak Proja Party had established itself in Muslim Bengal, Rahmat Ali spoke of two other Muslim states—Bang-e-Islam in Bengal and Osmanistan in Hyderabad (Deccan). These states, however, were to be quite separate from Pakistan as well as India.

The British brought in constitutional reforms in 1935. They introduced a system of provincial elections. As a result the Political Parties started to think in terms of the provincial elections. The Muslim League, which had always been aimed at the educated elite, now had to drum up support among the Muslim masses. Remembering the popularity of the Khilafat movement, they campaigned on a religious basis, and produced a number of hairy and unwashed Mullahs as champions of their cause. The Krishak Proja Party in Bengal, however, won most of the Muslim votes there without recourse to such methods. Fazlul Huq's main platform was the abolition of *zamindari* and the betterment of the Muslim peasantry. Unfortunately, when he came into office as Chief Minister of Bengal, in 1938, he was the head of a coalition cabinet which included the Muslim League. The League representatives in his cabinet included rich landlords who were utterly opposed, for obvious reasons, to the abolition of *zamindari.* Fazlul Huq compromised. To hang on to power at the head of the coalition, he was compelled to align himself with the League at the national level. The League, which in actual fact was composed of scattered and opposed elements, now took on, misleadingly, the appearance of a monolith. It seemed to speak for the mass of Indian Muslims, though as later events were to prove, it did not necessarily do so.

Between 1937 and 1940, a tremendous change took place in the attitude of Jinnah himself. He had previously been described as the architect of Hindu-Muslim unity, but from 1937 onward, he became the force that drove the Muslims towards the dream of Pakistan, which in 1933, when Rahmat Ali first proposed it, had been dismissed by no less a person than Zafrullah Khan, the future Pakistani Foreign Minister, as "a student's scheme. . .chimerical and impracticable." The Congress Ministeries which came into provincial power in 1937 were accused of doing down the Muslim population. It was the reaction of the Muslim masses to these accusations which triggered off Jinnah's new political parabola. As Richard Symonds writes, referring to the villagers of India: "Hindu and Muslim dwelt in the same village, but in different parts of it. The Muslim did not eat with the Hindu or marry into his family. His food, his dress, his name, and often his profession, were distinctly Muslim, and when he died he was buried, whereas the Hindu was burned." As long as the Muslim peasant could remember, adds Symonds, the British had judged and administered the people. But with the Congress—"however hard Congress leaders tried to make it otherwise, [it] was an overwhelmingly Hindu organisation"—in positions of administrative authority, the villager, "when he was told by League organisers of the wrongs which his Muslim brethren were suffering under Congress Raj...believed them.... If the British were leaving, there must be either Hindu rule or Muslim rule."

Jinnah in 1940 made a remark that resounded around Muslim India. "No power on earth," he said, "can prevent Pakistan." Symonds remarks, "...as we see now from the elections of 1946, the whole League leadership could not have checked the impetus to Pakistan even had it wished. The historical importance of the period of the Congress Ministries lies not so much in...whether the Muslim grievances were great or...exaggerated, but in the fact that the overwhelming majority of Muslims believed them to exist.... The overwhelming majority of Indian Muslims felt themselves to be Muslims first and Indians only second." Jinnah was the spokesman of these inarticulate masses: he had assumed, with typical arrogance and elegance, the role of Quaid-I-Azam, "Great Leader": but, amidst the more articulate Muslim politicians, he did not play this role unopposed. Jinnah would not allow the British to bypass his authority and deal directly with Muslim provincial leaders. In 1941, Lord Linlithgow, the Viceroy, announced the constitution of a National Defence Council. Various Muslim Chief Ministers, including Fazlul Huq, were appointed members. Jinnah ordered them to resign. Fazlul Huq, enraged, resigned not only from the Defence Council but also from the Muslim League Council and Working Committee. He remarked: "The genius of the Bengali race revolts against autocracy.... I could not... help protesting against the autocracy of a single individual." The Muslim League Ministers of his Cabinet resigned *en masse*. Jinnah issued a statement which had an echo of Stalin in it, "Fazlul Huq... has for some time now been trying to stab the Muslim League in the back." Fazlul Huq formed a Progressive Coalition Party, which took over Government in Bengal in December 1941. The Muslim League was now part of the opposition. In 1943, Huq resigned as Chief Minister. He alleged that he was forced out by a Cabinet plot of Jinnah sympathizers. Jinnah retorted that Huq, during his period of rule, had harassed and persecuted the Muslims of Bengal. "[He] is I am sorry to say a Mussulman."

Provincial elections took place in 1946. By this time, Jinnah knew that Pakistan must come soon, and the more territory it could lay claim to the better. It was therefore imperative that the League should be voted to power in Bengal. Led by Shahid Suhrawardy, it was. Far off in icy London the new Labour Government was trying desperately to assess the Indian situation coolly, and to solve it if possible, a difficult business with Jinnah demanding partition and both the Congress and the League determined to keep out of the proposed Interim Government. The Muslims felt that the Congress were receiving preferential treatment from the British. In protest, they declared a "Direct Action Day" for 16 August 1946. The results, all over India, were riots, nowhere more violent than in Bengal. 4,000 were killed in Calcutta, and a massacre of Hindus took place in the Noakhali district of East Bengal, where 50,000 people were rendered homeless.

This was by no means a new development. B. R. Ambedkar describes Indian history between 1920 and 1940 as a period of "civil war between Hindus and Muslims, interrupted by brief intervals of armed peace." He gives various statistics to prove this, and states that in Bengal alone between 1922 and 1927, 35,000 women were abducted, and that rape, murder, and the burning alive of enemies were common recreations for both communities. Partly this was due to the rooted opposition of each religion to the other : the Muslims of the east, for example, had reason to hate the Hindu *zamindars*—partly it was due to the anxiety of each community about its future in a free India which contained the other.

In October 1946, the Congress and the League combined uneasily to form an Interim Government in Delhi. There was a smell of blood in the air, and daily the situation became more unstable and unpredictable. Eventually, in June 1947, the new Viceroy, Lord Mountbatten, resolved on a decisive plan. In the Punjab, and in Bengal, the representatives of the Muslim and non-Muslim majority districts in the Provincial Assemblies were to vote separately on whether they wished to be parts of India or an autonomous Pakistan. The Sind Legislative Assembly were to vote on it as a whole. Adult male suffrage would decide the issue in the NWFP and Sylhet in Southern Assam. The representative institutions, in council, would determine the future of Baluchistan. Eventually, Sind, Baluchistan, the NWFP, and Sylhet voted for Pakistan. The western Punjab and eastern Bengal also chose Pakistan but the eastern Punjab and Western Bengal opted for India.

The new boundaries had to be drawn up: the smell of blood was more heavy than ever in the air. The British had originally intended to pull out in June 1948; but they were anxious to wash their hands, as soon as possible, of an intolerably difficult and treacherous situation, and retire with dignity. Mountbatten therefore, despite the pleas of both Hindu and Muslim leaders, announced that the British would leave on 15 August, 1947. On that date independence, partition, and the worst communal riots in history broke over the subcontinent.

Jinnah had previously described the country he accepted in 1947 as "a mutilated, motheaten, and truncated Pakistan." The East, without the industrial area of West Bengal, was of little value to him. As Richard Casey, the last Governor of Bengal, wrote before partition: "...If East Pakistan were to be confined to the Muslim majority areas of Bengal and Assam, I am not at all sure that it will represent the sort of Pakistan the Muslim League would want, particularly if it were to exclude Calcutta and the coal fields. The Muslim League is not seeking a sylvan retreat...." More particularly, it was not seeking a sylvan retreat separated from the 31,000 square miles of West Pakistan by the 1,000-mile breadth of India. The seperation would not have mattered so much if the industrial complex around Calcutta had come to Pakistan, but as it was there was little incentive for the West Pakistanis to interest themselves in their Bengali brothers.

Begum Ikramullah describes an incident that took place when the first Pakistan Parliament was convened in Karachi. "...A member from Bengal got up and moved a resolution suggesting that Constituent Assembly should meet once in Karachi and once in Dacca. Discussion followed... I supported the idea wholeheartedly.... The Prime Minister, Liaquat Ali Khan, got up to reply, and looking at me in mock horror said, 'Women never understand practical difficulty' to which I replied, 'If we had, we would not have got Pakistan today'." However, the Constituent Assembly, for the next nine years, never did meet in Dacca. The incident cited by the Begum was a small one, but significant, as a scrap of cloth on a stick can show which way the wind is blowing.

"We are *all* inferior in East Pakistan," Rahman cried. He raised his knees, and thumped them with both fists, beaming with happiness, like a large baby. His sarcasm had lost its bitterness. The friends who surrounded him knew exactly how to take it, he did not have to feel shy with me now, because the reaction of his friends would show me exactly how he meant what he said. The friends were all young, all about his own age: they sat about on chairs or on the floor, laughing, earnestly milking smoke from cigarettes, talking loudly, or displaying elaborate attention whenever I spoke. They had talked themselves out so far as poetry was concerned: they were mostly poets, some of them, I gathered, already well known—they had touched on the cyclone, and were now on politics. I had quickly discovered that all of them were communists. This was to be accepted and expected from young intellectuals in an underdeveloped country, but what did surprise me somewhat was that those of them who were not still at Dacca University were without exception Government servants.

Wahid, the host, plump and owlishly bespectacled, offered me a plate of vividly coloured sweets. "Please honour my house," he said, in what I presumed to be a direct translation of a Muslim formula of courtesy. "Rahman is not serious," he explained, while the others shouted, "No, no, this is only his playfulness," Wahid continued solemnly, "You know that at one time Dacca was the pearl of Bengal. Our handicrafts were famous — the muslin — yes, you know that? — but also it was the home of literature, famous writers and professors came from here. It was a very cultured city, and what is it now? It is a desert. The West does not care for us, we have declined. In 1947 there were more colleges in East Pakistan than in West Pakistan. So they built colleges in the West, but they have not built any new ones here. Yet our population is 54 per cent of all Pakistan. They are denying us opportunity." A young poet with El Greco hands and eyes, curled limply in a corner, commented: "Even with a college shortage, there are more unemployed graduates in

Dacca today than ever before. And if they were to travel to the West, they could find no work there, because they would be thought of as Bengalis, second class citizens, you see. We people here are the privileged class in Dacca, because we are the paid slaves of this Fascist regime." He laughed. "Look at the cyclone. So many of our people have died. They did not care in Islamabad that there was no flood control system in East Pakistan." Rahman chimed in. "Yes! Yes! Yahya will have said, the population density in East Bengal is so high, cyclone is good. Let the fellows die, that will solve the problem."

Someone threw a cushion at him. I said, "You don't seem to like West Pakistan." There was a yell of agreement. "Lots of people in India think," I said, "that a solution could be if East Bengal and West Bengal formed one autonomous state. . . ." I was interrupted by an almost audible silence. Eventually someone said, "You are not Hindu?" Rahman replied "No, no. He is British." Wahid said carefully, "You will not be offended? But we do not like Hindus." There were immediate and voluble explanations. "We have some Hindu friends in India. When it was possible to cross the border, we used to cross and meet them. Now of course it is not possible. But you see, those friends were individuals. As a people, the Hindus are evil." I asked how. "I was born in Calcutta," said the poet with El Greco hands. "I can remember what the Hindus did in 1947. Or at least, my parents told me what they did." One of the others obligingly simplified matters for me. "See, we speak and write Bengali. We revere Tagore, our great poet. We like some Hindus whom we know. But the Hindu people. . . . We do not like West Pakistan, but we hate Hindustan."

It wasn't surprising that they should. They were probably descended from people who had suffered exceedingly under

Hindu *zamindars*. Bred into their blood was a hatred of the Hindu as an oppressor, a race memory which they could not shake, and bred also into their blood was the intolerant Muslim hatred of the *kafir*, which they could not shake either. Moreover, they had been fed, since childhood, on substances nutritive to this hatred: fed by the Pakistani press, and Radio Pakistan. With the Indian borders sealed since 1965, they were not in a position to pursue a line of independent inquiry. They were intelligent, sensitive boys, but they disliked the West Pakistanis as, perhaps, their forefathers disliked the British, for their temporary assumption of power and overlordship, while the Hindus were their sworn and hereditary enemies. I had been told in Delhi, before I left for Dacca, that the East Pakistani aversion to West Pakistanis was intense, and that if the liberal Sheikh Mujibur Rehman won the elections in the East, his countrymen would embrace their Indian brothers with delight and love. It seemed fairly certain, at this moment, that the Awami League under Rehman would make a clean sweep of East Pakistan, but I felt that even when it did, my Delhi informants were in for a nasty shock.

Chapter 2

CYCLES OF CHAOS

In the first 10 years after its bloody birth, Pakistan had seven Prime Ministers and three political assassinations. Jinnah died, and the Muslim League subsided, rudderless, into the waves. The same obliterating sea that swamped it threw men up into power, then sucked them back, undermined a sequence of national and provincial Cabinets, and ate away the foundations of democracy. East Pakistan, a chaotic whirlpool, drew many to their doom. From the very start, it presented hazards that were impossible to overcome. They had been charted and foreseen, but no Pakistani politician, when he came to navigate them, could avert shipwreck.

"At Partition," Richard Symonds wrote, "the aspect of Pakistan which caused most scepticism among foreign observers was the viability of its eastern province.... Separated by a thousand miles from Western Pakistan, overpopulated and with no local industry, its commerce, administration and communications centred in the great Indian port of Calcutta, with a different language, dress, diet and way of life from the western provinces, it was rumoured to be likely very soon to secede quietly from its unnatural union with Pakistan, and merge once more with West Bengal as a unit of the Indian Dominion."

Some West Pakistanis today may devoutly wish it had done so. But the East Pakistanis in 1947 were extremely anxious to be free of what they thought of as Hindu rule. They had suffered under the *zamindars*, much more than the Muslims of the West. They ached for a Pakistani homeland, where they would cease to be a minority, and the concept of union with India was utterly alien to them. A speaker in the Pakistan National Assembly said in 1963, "...In Bengal, in particular during the election of 1946, when others [non-Bengali Muslims in other provinces] were not voting for Pakistan, 99 per cent of the Muslim population of Bengal did vote for the purpose of achieving Pakistan." It must be recalled, however, that this 99 per cent was mostly composed of poor and untravelled peasants, who knew nothing of West Pakistan or their fellow Muslims there. When in the fullness of time, after 1947, they became aware of the West Pakistani attitude towards them, they were surprised and annoyed. Their political leaders, the educated elite of the East, had been aware of the West Pakistani attitude before 1947: they had known it since 1912, when after the abolition of the 1905 partition, the Bengali cause went unsupported by the Muslims of the West. Shahid Suhrawardy, who had played an important part in the 1946 election victory, later said that things had happened too quickly, the rush towards partition had robbed the Bengali Muslim leaders of rationality.

The trouble in the East started immediately after Partition. The Muslim-controlled Government which had run the province since 1937 had set up a cadre of senior administrative officials. They had now to be transferred from the fleshpots of Calcutta to comparatively backward Dacca, and they did not like it. "Dacca," as Symonds wrote in 1950, "is a city of bamboo and corrugated iron, whereas Calcutta is proverbially a city of palaces." Colonel Mohammed Ahmed, who was attached to the staff of the first GOC of the eastern wing (this happened to be General Ayub Khan), wrote of this period: "The morale of the tough and rugged soldiers was low.... The discontent amongst the civilian officers and staff was at an alarming height." He adds: "Most of the senior officials had come from West Pakistan...feeling too unhappy to be discreet. They would complain in extremely bitter terms even in the presence of East Pakistanis, whom they blamed for all the misery and inconvenience. A thousand-mile wide gulf already existed geographically between the people of the two provinces. The termerity and impudence of the West Pakistanis was widening it emotionally much farther."

The West Pakistani attitude was stupid and insensitive, but not incomprehensible. The only bond between them and the Bengalis was Islam. The East Pakistanis were visibly poor and backward, and their country was underdeveloped, hot, and sticky. The renaissance of art in Bengal had been dominated by Hindus, so that even the local culture was alien to the western Muslims. Basically, the West Pakistanis in East Pakistan were a Middle Eastern people in a South East

Asian country. The place and its inhabitants seemed to them foreign. Their links with the East Pakistanis were no more than their links with any other far-away Muslim country like Indonesia. They behaved, in East Pakistan, as they might have behaved if posted as diplomats to Indonesia; not as countrymen, but as privileged visitors.

In 1948, shortly before Jinnah's death, a speaker in the Constituent Assembly said: "A feeling is growing among the East Pakistanis that Eastern Pakistan is being neglected and treated merely as a colony of West Pakistan." The Centre was in the west, and those responsible for rule at the Centre were selected bureaucrats and elected politicians. "So far as bureaucracy was concerned," writes D.N. Banerjee, "be it military or civil, East Pakistan was almost unrepresented, there being only one ICS officer from the eastern wing. East Pakistani politicians were also not an effective countervailing group against the dominance of Western bureaucrats backed by the charisma of the Quaid-e-Azam." The charisma vanished with Jinnah's death: the dominance of the West alone remained. As a sop to East Pakistan, Khwaja Nazimuddin, who was the Chief Minister there, was appointed Governor General in succession to the Quaid-e-Azam. He himself was succeeded as Chief Minister by Nurul Amin, whose supporters dominated the Muslim League in the East.

Nazimuddin, from the Centre, supported Nurul Amin, but dissension started in the League in Bengal. Several members left or were expelled. Maulana Bhashani, who before 1947 had been an obscurantist and vociferously religious leader, was one of these people. He set up the Awami Muslim League, which mounted a progressive platform. One of Bhashani's chief lieutenants was Mujibur Rehman. Similar splits took place in the western branch of the Muslim League, and the proliferation of new political parties increased the discords within the divided country. In December 1949, Hamidul Huq Chaudhry, Minister of Finance, Commerce, Labour, and Industry in East Pakistan, and described by Symonds as "perhaps the most energetic of the ministers," resigned his portfolio. He said he was disillusioned. The Centre proceeded against him, under the PRODA, or Public and Representative Offices Disqualification Act, whereby a recalcitrant politician could be debarred from all political office for a maximum period of 10 years. The reasons for Hamidul Huq Chaudhry's disillusionment came to light during the case that followed, and they represented the reasons for the disillusionment of all East Pakistanis toward their new homeland.

Among the nine reasons Chaudhry enumerated were: (*i*) he proposed to float a loan for his province in West Pakistan and the Centre forbade it; (*ii*) he drew up plans for mills, but the Constituent Assembly was made to pass legislature transferring industry to the jurisdiction of the Centre; (*iii*) he opposed Sales Tax being taken away from his province, but with no success; (*iv*) he pressed for a share of Income Tax for his province, but with no success; and (*v*) he asked permission to open a Commercial Bank in the province but the Government opened a National Bank. The list of failures is comprehensive. Chaudhry's frustration must have been complete. It was as though the West was deliberately blockading the advancement of the East.

On 16 October 1951, Liaquat Ali Khan, the first Prime Minister of Pakistan, who had ruled since 1947, was shot dead by an unknown assassin. Next day Nazimuddin succeeded him. For the next two years Pakistan plunged, like a sounding whale, deeper and deeper into chaos. In 1952 the Government attempted to impose Urdu, the national language, on Bengali East Pakistan. The result was riots, and Nazimuddin hastily conceded a defeat. His own countrymen of the East, however, now looked on him as a quisling. Moreover, since his advisers were mainly Bengali, he was not popular in the West either. In December 1952, a Basic Principles Committee published a report which envisaged a parity of seats between the Eastern and Western wings in the two houses of Parliament. This arrangement would have meant that the Eastern representatives would outnumber those from the divided West. The Muslims of the Punjab emitted a chauvinistic howl of protest. Dissatisfaction ran like wildfire through East and West alike, and the Governor General, Ghulam Mohammed, dismissed Nazimuddin and his Cabinet on 17 April 1953.

Nazimuddin's successor was another Bengali, Mohammed Ali of Bogra. Ghulam Mohammed, however, was the real wielder of power, but even he could not hold the gilded East in fee. The Awami League was now powerful in Bengal. Its platform was radical and leftist; and it demanded two concessions now close to the hearts of most East Pakistanis. It demanded complete regional autonomy, and it further demanded that Bengali be declared a State language. Shaheed Suhrawardy had now made the Awami League his political base, and with Bhashani and Mujibur Rehman shared the leadership. When elections were held in East Pakistan in early 1954, Nurul Amin and the Muslim League collapsed completely before a United Front spearheaded by the Awami League. Nurul Amin himself lost his seat to an 18-year-old student. The United Front claimed 223 seats: the Muslim League 10. The veteran Fazlul Huq, now head of the new KPP, took over as Chief Minister at the head of the United Front. East Pakistan now demanded complete autonomy except in foreign affairs, defence, and currency. "Save Pakistan from dissensions," pleaded the Frontier Gandhi, Khan Abdul Ghaffar Khan, "Bengal is our elder brother." The Central Government replied promptly by putting him in prison.

Scenes of violence flared up throughout the divided country. At the end of May, Fazlul Huq and some of his Ministers flew to Karachi to confer with the Central Government. They left for Dacca on the 30th, but before they reached it, Ghulam Mohammed issued a proclamation imposing Governor's Rule on East Pakistan. Next day Huq was placed under house arrest in Dacca. Suhrawardy, tactfully pleading illness, left in haste for Zurich, where he said his doctor was. Curiously enough, Bhashani also fell ill at the same time, though *his* medical adviser turned out to be in London.

With the Awami League leaders out of the way, Nazimuddin made an attempt to return to power with the Muslim League in East Pakistan. Huq and the KSP had been temporarily paralysed so Ghulam Mohammed decided to use Suhrawardy and his Awamis to quell Nazimuddin. Feroze Khan Noon, who was close to the Governor General, now flew to London. He followed the tradition of the Awami leaders in that he said he needed to see his doctor. From London, he flew to Zurich, where Suhrawardy met him. A statement, issued by Suhrawardy, asked the Governor General to dissolve the Constituent Assembly and frame a new Constitution after fresh elections from the existing provincial legislatures. These were held in June 1955, East and West having 40 members apiece.

A critical issue arose at this point. East Pakistan consisted of a solitary province, but the West contained three separate provinces, the Punjab, Sind, and NWFP. The pressures on the population of each of these provinces was different :

but while the East was one, the West was divided. What was known as the One Unit scheme was now proposed, by which the three Western provinces would be made into one. This would benefit the Punjab, which was the largest Western province, but it was of no benefit to anyone else. Sind and the NWFP were afraid that they would lose their identity and that their political voice would be muted. To East Pakistan the scheme was anathema. Though its population formed 56 per cent of the total population of Pakistan, this would be of no advantage. It would remain the Cinderella of the act, and there was no Prince Charming on the horizon. The issue, however, was postponed as a new political shift took place. In August 1955, the ailing Ghulam Mohammed created history of a sort : he became the first Pakistani politician who went on leave when he was actually ill. The Minister of the Interior, Iskander Mirza, took his place, and never relinquished it, until, three years later, he sank in the masterful wake of Ayub Khan.

Two days after Mirza's takeover, the Muslim League Parliamentary Party met, and elected Chaudhuri Mohammed Ali as its leader. So Mohammed Ali of Bogra was forced to resign. His supplanter pushed the One Unit scheme through the Constituent Assembly on 30 September. On 14 October, the new province of West Pakistan was created, and the interim Government of the province was headed by an Independent, Dr Khan Saheb. Meanwhile the Constituent Assembly was at work. It drew up a new Constitution, which was published on 8 January 1956. In March, Iskander Mirza became the first President of the Islamic Republic of Pakistan. He launched a new party, the Republicans, designed to ensure that he continued in office.

All this while the East fumed and flared. In September 1956 food riots in Dacca led to the Awami League being invited to form a Government in Bengal. This in turn led to a change at the Centre. Chaudhuri Mohammed Ali resigned, and Suhrawardy took over as head of a coalition of Awamis and Republicans. Maulana Bhashani and he clashed over his foreign policy and in 1957 Bhashani broke away from the Awami League and formed the National Awami Party. That October Suhrawardy lost his always precarious majority in the National Assembly and was forced to resign. Ibrahim Chundrigar, who replaced him as the result of a desperate coalition between the Muslim League and the Republicans, only lasted two months. Feroze Khan Noon took over from him in December.

There followed a 10-month period in which chaos became absolute. Governments rose and fell in the East. In the West, a political assassination took place in May 1958, when Dr Khan Saheb was murdered. On 23 September, the Deputy Speaker of the Provincial Assembly in Dacca was so seriously beaten up by opposition members that he died. Noon, at the head of a coalition of six very disparate parties, had to placate each one by appointing Ministers from it. The ludicrous result was that there were 26 Ministers in a National Assembly of 80 members : one member out of three was a Minister. Nor were these Ministers tractable colleagues. When Noon attempted to reshuffle their portfolios, the Finance Minister, a Republican, flatly refused to relinquish his post. Noon was forced to create a second Finance Ministry, which he called the Ministry of Commerce, and handed over to Hamidul Huq Chaudhry of the KPP. Meanwhile, the Awamis, none of whose members had received an important portfolio, quit the coalition with the political equivalent of a petulant toss of the head. Things could obviously not go on like this for much longer, and they didn't. On 7 October 1958, the President declared :

"I have decided that (1) the Constitution of 23 March 1956 be abrogated. (2) The Central and Provincial Governments are dismissed with immediate effect. (3) National Parliament and the Provincial Assemblies are dissolved. (4) All political parties are abolished. (5) Until alternate arrangements are made, Pakistan will come under martial law." The Supreme Commander, General Ayub Khan, was appointed the Chief Administrator of martial law. It was said that Mirza's proclamation had been prompted less by horror at the political state of the country than by fear that he would not, in a democratic election, be voted back to power.

But Mirza had chosen the wrong path, if he wanted to retain power. "This," said Ayub Khan on 10 October, "Is a two man regime. . . . If the President does not react to a situation . . . if something goes wrong . . . or he does something senseless, I shall act. . . ." On 27 October 1958, Mirza bowed out. It was now a one-man regime. Ayub said that the revolution was "based on consent, since there was no outcry or violent resistance." With the Army in power, backed by a military court apparatus empowered to sentence those who opposed it to death, this was small wonder. "Our ultimate aim," said Ayub, "is to restore democracy," but he was obviously in no hurry to do so. East Pakistan, meanwhile, was shut out in the cold. They had no senior representatives in the Army, their only power was in politics, and with political parties banned and elections forbidden, all they could do was to sit back uneasily and wait.

Some of them waited in prison. Bhashani was detained under the Security of Pakistan Act. Mujibur Rehman, Hamidul Huq Chaudhry, Abul Mansur Ahmed, and Abu Khalek were arrested under the East Pakistan Anti-Corruption Ordinance. A number of leading Dacca civil servants were also imprisoned. Ayub Khan was a simple soldier, though by no means as simple as he himself liked to make out. He was a believer in discipline, and the direct approach. No approach could have been more direct than Ayub's to the Eastern leaders.

On 26 October 1959, after a year of unquestioned power, Ayub made himself a Field Marshal. The next day, the anniversary of his dropping of Mirza, he announced the Basic Democracy Order. This enforced a five tiered system of local Government, with officials and elected representatives working together. There were 80,000 elected Basic Democrats, and in January 1960, they produced a dutiful vote of confidence in Ayub. If this could be considered an election, Ayub, on 17 February, became the first elected President of Pakistan. He promptly appointed an 11-man commission to draw up a new Constitution.

This misfired slightly. The Commission circulated a questionnaire asking for representative views. Thirteen representative East Pakistani organiations answered it by saying that they stood for parliamentary democracy. In the West, Chaudhari Mohammed Ali said the same. Ayub's retort was that the Chaudhari was "an old political hypocrite." Bluff and soldierly as always, he replied to the charge that he was a military dictator with the simple statement: "Bloody rubbish." When the Constitutional Commission extended the date limit for replies to its questionnaire, however, and thus prolonged the period during which open political debate was allowed to take place, Ayub warned them against "playing politics," and said that "the full force of Martial Law will be brought to bear against them." He also expressed the view that democracy "should not be rigid. It should not be a millstone round our neck." There was no danger of this, certainly, with Ayub around.

East Pakistan, however, was worrying him. He toured the province in May 1961 to inspect the damage caused by a recent cyclone, and during his trip promised to develop the eastern economy, and not to post West Pakistani officers to the East. In March the following year, he promulgated the new — and considerably amended — Constitution, based on the principle of a President. A tremendous publicity campaign was launched to boost the Constitution. It reached such a level that Ayub publicly asked forgiveness from Allah because "such praises are enough to upset one's mental equilibrium." This, however, was only in the West. Despite Ayub's 1961 promises, East Pakistan maintained what was described as "a sullen silence." The Director of the Bureau of National Reconstruction said cryptically that "the offence of propaganda against the Constitution was interpreted to include silence."

On 30 January 1962, Suhrawardy was arrested. This led to violent student demonstrations in East Pakistan, and Ayub, shaken, made concessions in the form of larger allocations to the East. He also inserted a clause in the Constitution, to the effect that parity between the provinces would be enforced in all spheres of the Central Government. That did not placate the Bengalis. After the Constitution had been brought into effect, the Dacca students struck for a day, demanding the restoration of democratic rights and a parliamentary Government in Pakistan. Political parties were still forbidden, but at the elections of April 1962 many of the former party leaders were voted into the National Assembly. "The tidy pattern of Basic Democracy devised by President Ayub and his colleagues," wrote the London *Times*, "has been disturbed at the first opportunity." On 8 June 1962, as the National Assemblymen were sworn in, Martial Law came to an end in Pakistan.

Ayub had now been nearly four years in power. He had ceased, in some respects, to be a soldier, and was a shrewder politician than before. But he was still blunt to a fault. When an East Pakistani politician demanded that the Constitution be amended, Ayub barked: "I would like to remind you that the recommencement of the constitutional process does not mean the restart of political life." But he revealed his newfound political acumen when he set about trying to split up the East Pakistani members with an offer of jobs for the boys.

However, he could see that the idea of a partyless National Assembly was "a miscalculation." He came to this conclusion after the turbulent East Pakistanis formed a united front of nine political leaders, who demanded an abolition of restrictions on their parties. Suhrawardy, who had been released from prison, allied himself with them. He was soon arrested and imprisoned once more, but the revival of political parties was well on its way and Ayub had started to think of forming one himself, an army to help him fight on two fronts. As the London *Times* wrote in 1963, "The politics of Pakistan have at present two dominant themes: an attempt in the name of democracy to wrest power from the executive and return it to the legislature, and the insistence in East Pakistan that this wing of the country receive an equal share in all economic benefits distributed by the Government." The two themes, in Bengal, were interlinked, for a dictatorship in the West threatened the very existence of the East.

A new party was formed, called, ironically, the Muslim League, and Ayub allied himself with it. Meanwhile, the death of Suhrawardy caused a brief lull in pressure from the East Pakistanis. They were leaderless for a little, but the lull did not last. By 1964 the Eastern leaders, notably Mujibur Rehman, were preaching separatism, and in March the revived Awami League demanded full autonomy for the East, and adult franchise. They got neither. In January 1965 the 80,000 Basic Democrats who comprised the electoral college voted Ayub back to the Presidency. (The total population of the two Pakistans, it must be recalled, exceeded 100 million people.) Of the 80,000 who did vote, nearly 50,000 were for Ayub. He won easily in the West, but had to fight in the East, where he defeated his opponent, Jinnah's sister Fatima, by 21,012 votes to 18,434. The *New York Times* wrote: "It may seem strange to Westerners to see a nation vote for virtual dictatorship over the possibility of internal liberty...but this was understandable in Pakistan...."

Later in the year, Pakistan went to war with India. The culmination of the 22-day conflict was a peace treaty drawn up at Tashkent. The West opposed the treaty, but the East did not. While the West still grumbled over Tashkent, Mujibur Rehman issued a six-point demand to Ayub, the heart of which was a demand for a federal constitution which would give the federal Government control of nothing but foreign affairs and defence. There followed a period of chaos and near revolt—Ayub, indeed, spoke of a Civil War—in the East. Rehman was imprisoned. Riots and gunfire shook the province.

The uproar died down, but the sullen embers remained. Ayub, tried by years of strain, fell ill in 1968. The next year the muted embers of revolt in the East flared to life. On 14 February 1969, after widespread disturbances in both wings of the country, the army was called out. Mujibur Rehman was released and attended a Round Table Conference between Ayub and the opposition on 12 March. Ayub, tired and ill, submitted to demands for direct elections based on adult franchise and a parliamentary system. The question of autonomy and representation on the basis of population was postponed. This caused the East Pakistanis, represented by Awami League, to break away from the rest of the opposition. By March a situation of anarchy had developed in East Pakistan, with arson and murder widespread. On the 25th of the month, Ayub, like Iskander Mirz nearly 11 years before, announced his abdication. He handed over power to the Commander in Chief of the Army, General Agha Mohammed Yahya Khan. In a statement made over Radio Pakistan on 26 March, Yahya said he had no ambition other than the creation of conditions conducive to the establishment of a constitutional government. He spoke of a sound and honest administration being a prerequisite for the smooth transfer of power to the representatives of the people, elected freely and impartially on the basis of adult franchise. At the moment, he said, "...normal law enforcing methods have become totally ineffective and have almost completely broken down. Serious damage to life and property has occurred and a state of panic has paralysed the life of the nation. Production has gone down to a dangerously low level and the economy ... has suffered an unprecedented setback. Strikes and violence have become a daily routine...." In view of all this, he concluded, Pakistan was to be yoked once more to the shafts of martial law.

Yahya, unlike Ayub, was comparatively quick to schedule elections. He assumed a slightly pathetic posture when he explained why, and declared that he was not politically minded and yearned to return to the simple life of the Army. There were, however, other reasons which forced him to be brisk, for East Pakistan was still astir, and could obviously not be contained without open conflict or elections. Polling, it was announced, would take place on 7 December 1970.

After the National Assembly had been voted into power, yet another new Constitution was to be drawn up. Yahya cautiously stipulated that it would not actually pass into law unless he approved it, but otherwise he adopted a democratically avuncular air about the whole business. Martial law had been suspended, and the campaigning parties worked hard in East and West. There were to be 313 seats in the National Assembly, 13 of which were nominated seats for women. Of the remaining 300, 162 were in the East, which contained 56 per cent of the Pakistani population, and 138 in the West, which though physically more than five times the size of its underprivileged sister, has a much lower population density.

The present nature of Pakistan was illustrated by the nature of the elections. No single party, like the old Muslim League, was powerful in both wings of the country. The Awamis of the East were a purely regional party, and part of their programme, indeed, called for provincial autonomy. They dominated the area in which they operated. The People's Party, led by the fiery and exhibitionistic Z. A. Bhutto, was powerful in the West, but had no influence whatever in the East. It was perfectly clear who the winner in each wing of the country would be. The only point of debate was whether Rehman in the East or Bhutto in the West would sweep up enough seats to dominate the National Assembly.

On 12 November, less than a month before the election, the cyclone came. It ravaged and slaughtered on the coasts of the East, and drew back from its own devastation to return seaward. It was the most immense natural disaster the world had seen in the 20th century, and it wrecked, not only hundreds of thousands of lives, but the aspirations of Bhutto. For in some peculiar way, the East Pakistanis seemed to blame the West for the cyclone. Logic had little to do with it. It was an emotional reaction born of years of Western dominance.

But, illogical though this feeling was, it had a political effect. Bhashani, 89-years-old and apparently indestructible, is the head of the NAP (National Awami Party); he declared that in deference to the memory of those killed in the cyclone, his party would not contest the elections but would occupy itself with relief work. This removed an obstacle from Mujibur Rehman's path, since Bhashani also advocates autonomy for East Pakistan, and the idea of autonomy now rang imperative bells in the ears of many voters anxious only to escape the toils of the West. Nearly the whole Eastern electorate, therefore, shifted over to Rehman. Observers in India drew a sigh of relief. Bhutto is a friend of the Chinese, he wants Kashmir and would probably be ready to start a war on the issue: Mujib is a dove by contrast, and advocates trade with India. But there are always unpredictables to be contended with during an election. The foreign observers could still only wait and see.

Dacca, even in December, was stiffingly hot by day. Drab and dusty, the city sprawled round its wide yellow river in a posture of exhaustion. A little wind off the water rustled the large triangular sails of the boats that lay at anchor by the quayside. Further out in the river were a couple of rusty steamers, the crew asleep in the patchy shade of the decks. Wahid, somnolent as a plump owl under the heavy sun, stood by me, blinking out across the river. In the sheds beside which we stood, there were other sleepers, bearded and saronged, with fizzing flies thick on their hands and faces. "This was such an important port, once," Wahid said, rousing himself. "Ships must have come here from Egypt and the West, to buy our muslin. It must have been busy." He gestured vaguely around. "Look what it has come to. My whole country reminds me of this. As though a great king had gone away, and everybody is asleep, to pass the time till he returns." Had he now returned, I inquired, in the person of Sheikh Mujib? Wahid shook his head. "No, no, he has not come yet. Mujib is only a messenger he has sent, to tell us he will come. And when he comes, he will come from the north."

"And his name," I said, knowing Wahid's politics, "will be Mao." He laughed. "I do not know his name. But I would vote for him if he were a candidate on the 7th. Since he is not, I will vote for Sheikh Mujib. Who else is there?" He mopped his brow with an appropriately red bandanna. "It is too hot," he said. "Let us take refreshment." We wandered away from the quayside into an alley of rickety tenements, the dome of a mosque visible in the distance. This was the old part of the city. We stopped at a kiosk which sold cold drinks. The owner was listening to the radio, which was emitting a stream of Bengali. "The Hindustani troops," Wahid explained, "have invaded a Pakistani enclave on our borders and kidnapped 300 people." I said it sounded to me very unlikely. Wahid shrugged. "Maybe. But I have no reason not to believe it. You ask this man, the owner, he also has no reason not to believe it. After all, we found out what the Hindus were like before partition."

We finished our drinks and retraced our steps to where we had left the taxi. "You know," said Wahid, "I am not a believer. But in this part of the city, my home, I always feel comfortable. Muslims have always lived here, the place is Muslim to the roots." All round us, I saw what he meant. The noisy, excitable crowd, with its bearded men, its occasional *burkha*-swathed women, looked remarkably fitted to its surroundings, to the lazy river behind us, the mosque in front, the still sun overhead. "It's an old city," Wahid said. "It's not only Muslim, but Bengali. Tagore came from Dacca, you know. His poetry is about this country, not about West Bengal. You know that really speaking we are the true Bengalis? As long ago as 200 A.D., our country was called Vanga, that is the ancient form of Bengal. We are a very old race."

We drove into the new part of the city. This is a recent development, and looks rather like parts of Delhi. Tall, gimcrack office blocks sprout out of the powdery dust, and somehow look unplanned, set down at random: there are inexplicable empty areas between them, where parked cars stand, lonely as though abandoned in the Sahara. More or less opposite the Governor's handsome white residence, which stands back at the end of a long driveway, is a square, utilitarian building which I took to be a factory. Wahid shook with laughter. "Sacrilege!" he said. "Sacrilege! That is the new mosque. The architecture of Muslim Bengal is unique, there is nothing else like it in the world. Sir Mortimer Wheeler has said so. But the new mosque in the capital of Muslim Bengal looks — yes, like a factory. But of course it is fashionable architecture. No doubt there are mosques like that all over West Pakistan."

All the new buildings in Dacca, not merely the mosque, seem strangely separate from the rest of the city. Dacca as a whole made me feel that it had become permanently fixed at some past point of time, where its rhythms matched those of its river. There was a ghostly quality to the old city, populous though it is. Its undeveloped quality made the tall new office blocks, not far away, look even more out of place, like suited Europeans in a *souk*.

The taxi took us back to the Intercontinental. I asked Wahid to come in for a drink. "Thank you, no," he said. "My wife will have cooked my lunch." I was startled by this. I had been in his tenement flat for nearly two hours the night before, and I had not once set eyes on his wife. This I would have understood in an orthodox Muslim household, but Wahid and his friends were the opposite of orthodox. Moreover, though he had talked a lot about his past life and ambitions, he had never mentioned his wife. "I have two children," Wahid told me now. He seemed rather ashamed of this. Then he hailed a cycle rickshaw and went off.

This mysterious quality about the young Dacca intellectuals struck me whenever I met them. They would be overwhelmingly frank about some areas of their lives, and entirely conceal others, for no apparent reason. Similarly, while most of them were Communists, and dismissed nearly all the ideas and statements of the Pakistani Establishment as reactionary rubbish, they believed, or seemed to believe, all that the Establishment said about India. They believed, for example, that the Indian Muslims were intensely persecuted, and led a miserable life. The Jan Sangh and the R.S.S., they felt, were dominant in Indian politics, and the Ahmedabad riots of 1969 had underlined that dominance.

"There are 60 million Muslims in India," I pointed out to Rahman one day. "That makes about 10 per cent of the total population of India. There's no exodus of Indian Muslims into Pakistan. Surely, if they were persecuted, there would be." Rahman looked a little perplexed. "In 1949," I added, "there were 12 million Hindus in East Pakistan. That made 25 per cent of the population. How many are there now? All over Pakistan, the percentage of caste Hindus is 8 per cent and there are 5.8 per cent Scheduled Caste Hindus. All the rest have fled to India. There is no Hindu party in Pakistan, but there are Muslim parties in India. So you really have to work out who persecuted who, and which country it happened in."

Rahman and the others listened to me, because they knew that I was an expatriate, and held a British passport, and had no vested interest in India. But they were not convinced. They had no arguments against the facts, but they simply would not accept them. Had they been less intelligent, I would have felt less surprise. But they were intelligent, and their politics had made them disbelieve the information of the Pakistani press and radio on everything except, it seemed, India. It was not that they had been brainwashed in this respect: it seemed to be a deep compulsion in their blood which made them dislike Hindus and their country. Yet, like the Awami League, they accepted the idea that trade with India was desirable. "But trade does not mean friendship," they repeated, and grimly shook their heads.

International airports and hotels are enveloped in a sleazy kind of romance. They are gilded beehives, filled with the buzz of complex and separate lives. Many unreadable best sellers have been written about these public places where private people collect, and I thought the Intercontinental, in the weeks between the cyclone and the election, would have provided surefire material for another. The characters were ready to hand: if the author wanted a model for a tough, seasoned foreign correspondent, there were dozens available; there were American Air Force officers equipped for instant transference to prose, all with cold blue eyes and desert boots; there were earnest Red Cross and A.I.D. workers, and loquacious Pakistani liaison men. Miles to the south, down the broad tropical river, the night hovered over the wrecked offshore islands and coastal areas, where thousands of people recently deprived of their families and homes crouched under improvised shelters against the cold. Pathos, comedy, tragedy, irony, were all readymixed in the situation. Sex and violence could doubtless be located if required. Perhaps that was why I spent as little time in the hotel as possible. Living in it was like being trapped between the covers of some unwieldy American bestseller.

So I was delighted when the Press Officer told me that a Central Minister had decided to throw a party for the foreign correspondents. It ensured an evening free of the Intercontinental, for one thing. For another, the only people I had so far met in Dacca had been minor officials and my young intellectuals, and I was anxious to see how the other half lived. The Press Officer assured me that all the notables of the city would be present at the reception. "A glittering occasion," he said it would be. It would certainly, I reflected, be a crowded occasion, for there were 142 foreign correspondents in the Intercontinental alone, and others dotted around different hotels in the city. The reception was to be at another large hotel. That evening there was a holiday atmosphere in the Intercontinental: correspondents who had tapped at their typewriters for days lolled back in the bar, dressed in their best, and waited, like small boys, till it was time for the party.

The reception was like any similar reception in Delhi. The Pakistanis present were sleek and westernized, and several of the men wore evening dress, which rather put us to shame. Curiously, it was immediately possible to tell which of them came from the West. The West Pakistanis were taller and had fairer skins, and their conversation was impeccable party conversation. I became entangled in a long, very dull dialogue about Oxford with one of them, who had been there. When I finally extricated myself, I went over to a corner full of East Pakistanis. They were talking about the elections, and

I discovered that they were all local business people. They were wealthy men, but despite Mujibur Rehman's socialist platform, they were for him. "My ancestors," said one of them, "came from the same peasant class of society that Mujib is trying to uplift. There's no doubt that the mass of people here are backward: unless they are uplifted, East Pakistan will never be able to function as an autonomous unit." There was no question, he added, of the East breaking away from the West completely. "We cannot do it. We need each other. Perhaps the West needs us more than we need the West, but we could both function together if both were autonomous. In the present state of affairs, we are dependent on the West, because the West thinks that if they keep us dependent, we cannot break away. The East, if it is kept dependent, will be forced to break away. This imposed dependence has been the chief cause of trouble in Pakistan. If the East is made autonomous in everything except foreign affairs, defence, and currency, which is all Mujib wants, Pakistan would work more smoothly as a whole than in the present situation, where they have chained us to their wheels. The question of our physical division does not really arise. We may be thousand miles from the West, but look at India. Their problem, in a sense, is the same. Take a man in Amritsar. He has probably never been to Madras in his life, he knows nothing about the conditions there, and it is a thousand miles away, but he still feels the Madrasi and he belong to the same country." This seemed to me a slightly optimistic view, but I did not pursue it, since I was interested in what they felt about India. "Of course," one of them said, "trade with India is desirable. It would lead to more friendly relations. War is a drain on our resources: why should we fight India? The temper of our masses at present would probably not allow us to be cordial with them, but at least we can open our frontiers and trade, and perhaps slowly come to be friends. It's the West which has opposed trade with the Indians, and imposed a block on it: the loss of trade doesn't affect them, it affects us."

Another man remarked : "You have heard of the 22 families? They are the richest people in Pakistan, and they have financed every Government which has lasted. They have been in a position to impose their own views on every administration. They are the biggest evil in the country. They have dictated the treatment of East Pakistan—because they are all from the Western sector—and the policy vis-a-vis India is dictated by them also. They must be rooted out, and Mujib is the man to do it. They won't be able to corrupt Mujib as they corrupted Ayub." The first man who had spoken chimed in. "You know, for all Ayub's faults, we must be grateful to him. At the time he came, the country was in such a bad state that its survival seemed impossible. He was a dictator, certainly, but he made it possible for Pakistan to continue."

I noticed, when they spoke of Pakistan, however, there was a certain inflection in their voices, a careful noncommittal sound. It puzzled me, and I delved a little, trying to find what their concept of the country was. Eventually, a man who had not spoken before, a short tubby little person with inappropriate military whiskers, said, "We are Bengalis here, you know. Bengali culture has deep roots, and it's our culture. So, when we fly to the West, for example, we feel we are in some other country. Yet our friend here earlier said the problem in India is the same, and that is true. I would say that both here and in India the country is founded in a partnership between the different cultures and people, exactly like a business partnership. The partners need not like each other, but for the survival of the partnership they must work together. I am a businessman, and I see East Pakistan and West Pakistan as partners in a company which is united Pakistan. We will see that the company prospers, since it is our livelihood, but we demand an equal partnership. It is very simple," he said, beaming at me. "Now let us take some food."

The businessmen at the party interested me. They were intelligent, articulate people, and members of the Establishment, and they supported Mujib. They did so largely because they were East Pakistanis, and he stood for the welfare and autonomy of East Pakistan. The same reason motivated the masses, who also supported Mujib. In so far as his policies outside East Pakistan went, however, there was a difference. The businessmen, insulated by wealth and position, were not afraid of India, and they favoured Mujib's proposal to renew trade between the countries and establish friendly relations. The masses, however, were hostile and wary of a nation that was predominantly Hindu, and one foresaw that Mujib would have trouble in implementing this aspect of his foreign policy, if he ever received the opportunity to do so.

There was no doubt, however, about the imminence of his victory in December. The opposition parties scrawled their election symbols up on the walls and tree trunks all over Dacca, but these were mere motions, the gestures of so many Canutes against the Awami tide. Rahman came to the hotel to escort me to a campaign procession for Mujib. "This will be fun," he sparkled, full of the euphoria that touched and left him with such mercurial rapidity. "It will be very much fun."

Mujib is the darling of Dacca. When he was released from prison in February 1969 to attend the Round Table Conference at Rawalpindi, thousands of his admirers flocked to his house, laden with salutatory flowers. "People of all ages and from all stratas of society came out into the streets, many dancing and raising slogans welcoming him. Crackers were fired to celebrate the occasion. The same evening an estimated 100,000 people gathered at the Paltan Maidan to extend a public welcome to him. Welcome arches were hurriedly erected. Black flags which had been flying from private and public buildings, in connection with the people's protest movements, were replaced by the national flag. A large portrait of Sheikh Mujibur Rehman was taken out in procession." The present occasion, though nothing like so spontaneous, was very nearly as widespread.

We drove into a distant part of the city. There were crowds out everywhere, some with banners, all as euphoric and noisy as Rahman. He escorted me to a friend's room, on the top floor of a tenement which overlooked the route of the procession. Across the street, other people were hanging out of windows, emitting anticipatory yells and exchanging badinage with friends in the street below. A dense crowd milled in the street: policemen with long sticks hustled them off the road, but they swelled back once more, in disorderly overflow, and small comedies were enacted as the policemen, booed by the spectators at the windows, chased adventurous individuals back into the crowd. To anyone new to Asia it would have looked like a riot, but it was not. It was merely a *tamasha,* the way things are done in the East.

The procession arrived, and a blare of film music came with it. This was momentarily drowned in a fullthroated roar of Mujibur's name. The lead cars of the procession were festooned with balloons and streamers, and the crowd shouting the name of Mujib fought to pluck them off. Banners waved above it all. There was an inextricable entanglement of the procession and the people, but even above the thunder of the crowd, and the scrum of men and banners, and the occasional crackle of fireworks, the ubiquitous film music wailed, and a few stray, vividly tinted balloons floated up to be snatched at by the watchers in the windows. Eventually, with the labour of a dying caterpillar trying to shake off a horde of ants, the procession reached the corner of the street, and, leaving an inharmonious echo of music behind it, oozed away out of sight.

Mujib was later criticized for the indecorous frivolity of this procession. Film music, balloons, and streamers, said the press, were not suitable as accompaniments to a political rally, so short a time after the horror of the cyclone. The crowds who packed the streets, however, raised no query. They were the people who would vote for Mujib, after all, and they had laughed over the balloons and the music : though if the cars had been draped in black, and the music had been funereal, they would doubtless have wept with equal intensity.

This procession was quite obviously aimed to please the illiterate and inarticulate supporters who would vote Mujib to power. The Awami League was taking no chances : they were keeping their name before the eyes of the voters : if they couldn't yet provide the mob with bread, they could at least provide it with circuses.

Chapter 3

DEATH BY WATER

Geography as well as history has been unkind to East Pakistan. The Bay of Bengal is shaped like a funnel, at the apex of which the country squats amid its swamps. The disturbances that take place in the Indian Ocean, particularly in May and October, when the rise and shift northward of hot air from the equator occur, tend to be forced into the mouth of the funnel. As the landwalls on either side close in, these disturbances are squeezed into a narrower and narrower compass, till eventually they explode at the apex of the funnel, on the offshore islands and coasts of East Pakistan. For centuries now, as overland invaders tramped into Bengal from the north and west, cyclones have walked up on heavy feet across the sea from the south, smashing into the coastal areas, taking people, houses, and animals in their iron hands, shredding them, and throwing them away. In 1876, one such invader from the sea killed 300,000 people. This is recorded, but it is very likely that there were worse cyclones before that, which went unrecorded, or of which the records have been lost. There was no protection, no means to chart or ward off cyclones, and for centuries the East Bengalis accepted and suffered their frequent unwanted visits.

Over the last 80 years, 42 cyclones have hit the area. The fact that it is prone to these phenomena is no secret, but nobody seems to have thought of doing very much about it. Since 1945, however, it has become possible to chart the course of cyclones, at first with scout planes, but more recently with radar and through satellites. In 1965, when East Pakistan was swept by a really serious cyclone, the Government promised to build embankments and dykes and implement an efficient flood control system. It set up a radar station at Cox's Bazar, near Chittagong, to spot and chart the cyclones as they came, and issue warnings to the population affected, but it forgot, conveniently, about the flood control system. The local peasants, of their own volition, constructed low embankments of packed mud, none more than 20 feet high, to protect themselves. These embankments were fairly effective so far as minor disturbances were concerned, but nothing large hit East Pakistan between 1965 and late 1970.

The historic cyclone of 12 November 1970 was the second to hit the coast within three weeks. On 23 October, its precursor, a noisy but smallish John the Baptist, had swept across the offshore islands to the mainland. The radar station at Cox's Bazar had spotted it some hours before its arrival, had clanged its alarm bell, and sent out reports and warnings, and though the damage to property and life was not fully assessed, it was relatively inconsiderable. But as the hubbub of the first cyclone died away, hundreds of miles away in the Malay Sea, unknown to any radar, its huge and fatal successor was slowly starting to build up.

Much about cyclones, as about cancer, is still unknown. But the genesis of a cyclone is now explicable. It starts over a low pressure area in tropical seas. The low pressure area, which initially forms and always remains the centre of the storm, is known as the eye. The main cause of all the trouble, it is a peaceful area of calm seas and clear sky, like some beautiful woman who has incited her lovers to brawl around her. As the beauty's interest in all the brouhaha decreases, the lovers quarrel more desperately: as the pressure in the eye falls, it is encircled by fighting winds, which whirl round and round it at a speed which can reach 200 m.p.h. This area of the cyclone, which is called the eye-wall, is the real killer. It may measure 50 miles across. It pushes up before it, as it comes, a jagged wall of water, 20 or 30 feet high.

The November cyclone heaved itself up, like some black monster, from the sea off the Andaman Islands. It lumbered ponderously up the funnel of the Bay of Bengal, at a speed of roughly 10 m.p.h. On 9 November, the American satellite Essa, floating high overhead, focused on the huge dark spiral as it moved over the waves, and sent instant warning flashes to Cox's Bazar. Meanwhile Indian cyclone experts were trying to keep track of the cyclone, afraid that it could veer, unpredictably as a missile out of control, southwest to Madras or north-west to Calcutta. It did not do so. Heavy and inevitable, it moved up the Bay of Bengal, and appeared on the radar screens of Cox's Bazar. Radio Pakistan immediately started to broadcast warnings to the areas likely to be hit.

The broad rivers of East Pakistan, once so essential to trade, pour into the Bay of Bengal. The Ganges and the Brahmaputra deposit immense quantities of silt in the sea, and this masses up into islands. The silt soil is highly arable, and the islands are therefore densely populated. Few of them, however, rise more than 20 feet above the sea. On the mainland shore facing them are other low lying, riparian districts, also densely populated. The cyclone would hit the offshore islands and then the coastal districts. It was all mapped out, by Essa and by the Pakistani radar. Radio Pakistan said so. What it failed to say, curiously enough, was that the cyclone coming was an immense one, and that the areas in question were in terrible danger.

The population of the islands and the coast heard the warnings, on new transistors or rickety old crystal sets. Unprotected as they were by any organized flood control system, their choices of salvation were limited : to climb to the tops of high houses and trees, and there try and ride the tempest out, or to flee for their lives. Most did neither. There were two reasons for this. Firstly, Radio Pakistan had cried Wolf three weeks before, when the cyclone of 23 October had struck. Not much damage had been done at that time. The radio had not emphasized the force of this new cyclone, so many people did not take the warning seriously. Secondly, even if the people were to flee, where were they to flee to? The population of the islands could only hope to reach the mainland coast, which was to be hit anyway. The coastal areas in many parts had no road or rail communication with the interior. An evacuation of the coast would have to take place along the rivers and waterways, and to be caught on a river in the midst of a cyclone of whatever size was obvious madness. The livestock of the peasants, anyway, could only be moved slowly, and were far too precious to be abandoned to wind and wave.

So the people stayed where they were. On the night of 12 November, most of them were asleep, under coarse blankets, in thatched straw huts on the islands and the coast. It was high tide, with a rough, breaking sea, and rain falling. Over the sounds of rain and sea came a distant roar, swelling slowly to titanic dimensions, drowning all other sound. Then an immense and deafening wall of water towered across the waves. Nobody, obviously, took a tape and measured it, but some peasants have said it was fifty feet high. It thundered down on the offshore islands. Houses, people, animals, trees, vanished under it. As the wave roared on coastward, a blinding, driving, killing wind followed it over the wrecked islands.

Wave and wind now exploded on the mainland shore. They rushed onward, unchecked and uncheckable, cutting a swathe fifty miles across which all villages and living creatures were ploughed irresistibly under. At last the wave halted, and then came the new terror. All waves, once they have broken on land, recede into the sea. The titanic mass of water that had broken on the East Pakistan coast was in fact a wave, and it obeyed this principle. It swept back seaward the way it had come. The survivors of the villages in its forward path now found themselves enveloped in choking, muddy water as it returned. Thousands of people, together with dead cattle, the wreckage of houses and boats, and uprooted trees, were swept helplessly out into the open sea.

Communications in the coastal areas are limited at the best of time. The cyclone destroyed what there was of them completely. It was two days before Dacca discovered the extent of the catastrophe on the coast. Helicopters, humming and flickering above the disaster area, saw thousands of bodies, human and animal, afloat in creeks, waterways, and the sea. The damp and oppressive heat had already commenced its work of decomposition. There were survivors, however, even on the islands, though the extent of the destruction was such that all that remained on one heavily populated island was a dog. These survivors, enveloped by the stench of corpses, and foodless, through the still hot days, and frozen and sleepless through the chilly nights, were also in danger of death. Hunger and cold apart, the likelihood of an epidemic caused by the unburied dead was terrible and real.

The Pakistani Army was despatched to the flood zones to bury the dead and help those who still lived. Within hours, however, as the news was flashed around the world, assistance started to pour into the country. The U.S.A. and Britain flew in troops to help with the relief work, and lent helicopters and launches to implement it. Medicines, food, clothes, and money were donated by other countries. India offered her condolences and a crore of rupees, which were accepted. She also offered supplies, which were accepted too, on condition that the Indian relief lorries stopped at the East Pakistan border, and handed over their loads to local officials. Other countries which provided supplies sent their own men into Pakistan to see that they were properly distributed, and it was alleged that the local officials who received the Indian supplies were selling them on the black market.

Pakistan had twenty odd helicopters available for relief lifts. The U.S.A., rather stingily, only lent them eight. India, however, offered fifty helicopters to Pakistan, complete with crews. The Pakistanis refused to accept them. The presence of so many Indian Air Force personnel, they said, would be a security hazard to the country. It seemed a curious time to think of security, but strangely enough most Pakistanis I met in Dacca approved of the Government decision, on the *ut dona ferentes* principle. Even more strangely, most of the flood affected peasants approved of it too. They usually hadn't heard that the Government had refused an offer of Indian helicopters, but when they were told this, they said the Government was right. "We should not," said the headman of a village off the coast of Bhola, "accept help from the Hindus."

But Pakistan gladly accepted help from everyone else. The Red Cross sent relief workers: so did A.I.D.; F.A.O. sent experts and observers. Even quicker than the aid came the foreign correspondents, who swooped down on Dacca in rapid succession, with a loud flutter of typewriter keys, a whirr of Arriflexes and Eclairs. They made sure that the world knew all about the disaster, and the world responded. There is nothing like a natural calamity to fill people with an undemanding pity (partly caused by relief that it did not happen to them) and a comfortably temporary sense of the brotherhood of nations. East Pakistan, for a short space of time, made daily headlines in every paper in the world.

Yahya Khan flew over from the western sector, spruce, military, and full of the correct responses to the situation, that is up to a point. Beyond that point he betrayed himself. "We are in the hands of Allah," he remarked when pestered with questions as to why the Government had not attempted to provide a flood control system in the East, "Who can stop a cyclone?" He may have been speaking as a good Muslim, but he was certainly not speaking as a good leader. Another distinguished visitor was the Pope, who stopped over in Dacca on his way to Manila (where he was nearly assassinated). He donated a substantial sum towards flood relief — which the Vatican could certainly afford—posed for the cameras of the press, and wafted a papal blessing to the people of the flood areas, who probably did not know who he was, would have disapproved if they had known, and, preoccupied as they were with the problem of survival, did not care anyway. It was good publicity, however, for the Pope and for the Pakistani officials who welcomed him at the airport.

Meanwhile, the death statistics fluctuated from day to day, as in some eerie stock market. The Pakistani press, for example, by and large, estimated that there had been more deaths than the foreign press did. The foreign press figures, in turn, were far in excess of those produced by the Government. The night before I left for Bhola, I strolled over to the Cyclone Information desk in the hotel lobby for the latest press release. It said that 191,000 people had so far been ascertained to be dead. Five minutes later, in the bar, a Dacca pressman said, very seriously, that he was sure a million were dead. "That," he said with an air of pride, "makes our cyclone," his air was positively proprietary, "the worst natural calamity in history. When the Yangtse Kiang flooded in the 19th century, 900,000 Chinese died, but we have beaten that! Pakistan," he announced, "holds the world record," and I could detect no trace of irony in his large mild eyes. It was curious how little these statistics meant to anybody. The press men and relief workers who had visited the flood areas were visibly moved when they talked of individuals they had met, who had lost families and livelihoods; but they could accept the idea of several hundred thousand individuals dying without emotion, for it was impossible to think of so many as more than a statistic. The magnitude of the death figures surprised them: it did not move them.

"Such a disaster," said the Dacca pressman, smacking his lips over his beer, "has also a purifying effect on young people. All over the world, young people of all nationalities are thinking of the disaster and feeling pity. This will give them a more serious attitude towards life." He wiped the froth off his mouth with a crumpled handkerchief. At the table behind us, two skimpily skirted American blondes, neither of whom was much more than 16, sipped Cokes and talked to a couple of Phillipino youths. They were all, I gathered from their conversation, which was not muted, the children of relief workers. "It would be fun to have a picnic tomorrow," one of the girls said. The older of the two boys clicked a contradictory tongue. "All the pretty places around here," he reminded her, "are crapped up with floods and dead bodies and stuff like that." The young blonde shook her pretty head. "Hell, I forgot," she said. "What a nuisance. But, I guess, no picnic."

The village had been utterly shattered. Of its fifty huts, none remained. There were no cattle in the fields, and no children in the muddy street. In the bamboo shelters which the Americans had put up sat the survivors of the population, nearly all men. The trees around the village lay like corpses in the damp earth. Square cardboard boxes of food, clothes, and medicine had been piled up in one of the shelters, preparatory to being opened and distributed. The survivors did not look towards them.

The American chopper, in the distance, moved like a dragonfly, darting up through the still air, then hanging there as though transfixed by astonishment before it darted up a little further. It was off to drop supplies in another village. "They will need supplies," said Captain Tariq of the Pakistani Army, nodding towards the survivors, "for a long while yet. The crops were ready to be harvested, and now they are ruined." As the sound of the chopper receded a deep silence lay down in the bedraggled fields around us.

Bhola is a large district on the southern tip of East Pakistan, and statistically the worst hit part of the disaster area. 119,000 people are estimated to have been killed here, either during the cyclone or by its aftermaths of starvation and cold. The Government figures, here as elsewhere, are queried by everyone, even the officials whose duty it is to issue them to the press. Flying over the district, one has the impression of utter devastation. The green landscape is blurred and smeary with ochre water, there are hardly any trees left standing, and the small relief encampments can be seen amidst the wreckage of large villages. The calculation of the death toll is not based on any census, since no proper census exists for the area, nor does it take into account the large numbers of people who have simply disappeared, and whose bodies have not been found. They may now be at the bottom of the sea, but until they turn up, or some kind of authoritative statistics are available, the Government presumes them to be alive. People, in fact, were picked up in the open sea, clinging to driftwood, and in one case to a dead and bloated buffalo, or perched on rickety rafts, some days after the cyclone, but there were very few such cases. Most of those who cannot be found are almost certainly dead. The Government count of dead in East Pakistan as a whole is similarly fallible. The estimate when I was in Bhola was, as I have said, 191,000 dead. But apart from the missing people, and leaving aside the lack of accurate census figures, nobody seems to have counted the numbers of fishermen and boatmen who were on the sea or the waterways when the cyclone came, or the numbers of people who travel from as far north as Sylhet, at harvest time, to reap crops in some of the offshore islands. Beside the islands with a known and permanent population, there are 14 which have a floating population resident only during the harvest. "The real number of dead," said my Army friend, "is probably anything from 300,000 to 500,000." He was a young man, erect and slender, but at the moment he looked terribly tired.

In one of the shelters a man squatted alone. He was thickset and short, his tufty beard dyed patchily red. His clothes were torn and stained, but a new blanket was draped around his shoulders. He grasped its corners, mechanically, with both hands, as though for support. His face and eyes contained absolutely no expression, and even his body was listless. When, with the Captain as translator, I spoke to him, he answered dully and automatically, as though a ventriloquist were speaking through him. "I had eight children and a wife," he said. "Four of my children were boys, four were girls. The eldest, my daughter, was eleven, and the smallest, my son, six months old. When we heard the water and the wind, I tried to think what to do. But there was no time. It seemed only a moment before we were in the water. My wife was carrying the baby. I caught hold of one of her hands. I caught hold of one of the children's hands. I told the children to hold on to each other. In this way, whether we lived or died, I thought we would not be separated.

"The force of the wave was terrible. I tried to brace my feet and stand fast, but it lifted me as I would lift one of my children. It was so dark I could not see, and because of the noise of the wind I could not hear. But I still had hold of my wife's hand and one of my children's hands. Then I was lifted and thrown hard against something solid.

For a moment I was stunned. When I recovered I had my arms around a tree. But my hands were empty. God forgive me, I did not try to look for my family, but clung to the tree till daybreak. I have not seen my wife or my children since."

This man, Hussein, I was told, had wandered around for three days after the cyclone, in a state of madness, looking for his family. Then, when the Pakistani Army and the Americans arrived, he had briefly returned to his normal self. He thought that they would be able to find his wife and children, since he could not believe they were dead. After some days he abandoned hope, and lapsed into the kind of stupor in which I found him. But it struck me that most of the survivors of the village were in the same state of shock. One could hardly expect them to smile, but they did not even look sad. They simply looked numb, expressionless. It was the same in all the villages I visited. The quiet survivors answered questions in dull and uninflected voices. Some had lost husbands, some wives: nearly all had lost children. When the Pakistani Army buried the corpses in Bhola, they were considerably hampered by the fact that the survivors kept digging the graves up, looking for their dead. They had given that up now, and subsided into shock, though in one village a young man wept terrible dry tears, his head in his hands. "They do that sometimes," I was told.

The airport at Dacca became my operational headquarters. Behind the normal traffic area, a large sector devoted to relief operations had been cordoned off. A complex of tin sheds housed the offices of the Pakistani and foreign relief workers.

Truckfuls of the square cardboard cartons of supplies were unloaded here: relief planes landed with more. The relief workers, too, had their martyrs: a Swedish plane with a crew of four had recently been lost. Tough Pakistani soldiers watched as the supply cartons were unloaded and stacked. It was rumoured that some local officials were plundering the cartons and selling supplies on the black market, but it was not a very reliable rumour. Certainly everything looked efficient and ran smoothly, except for a gloomy man who sat over a perpetually dead telephone in a billowy khaki tent near the helipad. He was in charge of putting the press on choppers or planes for the flood areas, but his task was not easy, as one could tell from his face. I did not rely on him, but with Rahman as interpreter prowled the offices in the tin sheds, trying to find someone who would put me on a plane to Patuakhali. This was a very difficult place to reach, but one crisp pale dawn Rahman succeeded in fixing a ride for me. An amphibian was on its way there to pick up a Pakistani Minister who had been on an inspection tour of the flood areas. It was ready for take-off when my ride was confirmed, and already contained Captain Rehman, a cheerful moustached man with a pipe, who was the pilot. Captain Rizvi, a bright youngster with sideburns, and Colonel Shaukat, who was young for his rank and had helped fix the ride. There was room for one more, and I squeezed into the back with Shaukat.

As we roared down the runway, he lit a cigarette, smiled, and said, "Never been to Patuakhali yet. Always flown the other way so far. That's why I'm coming along. Always interesting to see new places, isn't it?" Rehman talked into a stuttery intercom as we tilted up steeply over the sprawl of Dacca, then switched it off, and looked back at me. "You realize, of course," he said, "you'll have to find your own way back from Patuakhali. There won't be room for the Minister and you." Since I already knew this, I was not as horrified as I might have been. Grinning, Shaukat said, "I hope there *is* another way to travel from Patuakhali. Godforsaken dump, from all I hear."

We flew at a low altitude over the emerald landscape, streaked with broad yellow waterways, to which I had now become accustomed. Small boats floated down the waterways, triangular sails raised to the wind: the flow of life had restarted after all the deaths. After roughly an hour, Rehman turned back to yell, "Fasten your belts. We're going in now." Below us lay a wide river, with a scatter of huts and houses on its left bank. The water was cluttered with fishing boats and launches. Rehman clicked his tongue in irritation, so loudly I could hear it above the roar of the engine. Then he flew in low on a trial run. We thundered in over a huddled village of boats, the fishermen ducking and shielding their heads with their arms, and pulled out and round on the wide sweep back, trees and huts twirling beneath the slewed belly of the plane. "O.K." said Rehman.

We curved round, and came back in over the boats. As we zoomed down over for the landing, Shaukat yelled a warning. Coming rapidly towards us, directly in our path, was a large launch. Our engine roared, Rehman threw himself on the stick, and we shot up, feet above the launch, and pulled round in a screaming vertical turn, back over the huts and trees and round once more to the river. Rehman is a fine pilot, who during the extreme crisis period had brought his amphibian down in places where nobody else had ever landed, but all this had obviously shaken him considerably. As we shot downward over the boats towards the now launchless river, Shaukat shouted, "Too high! You're coming in too high." He plunged forward for the stick, with which Rehman and Rizvi were now wrestling in obvious panic. The creased sunshot surface of the river hurtled up at us. Then we hit.

There was a sound like an explosion. Yellow water rose round us like a wall. Shaukat bellowed in my ear, "Can you swim?" I was about to say that I couldn't, when the water subsided slowly into spray. Shaukat slumped back in his seat, fumbled for his cigarettes and said, "No need, after all." We had come down nosefirst, but had righted ourselves somehow. As the spray settled, the world came back into place. The amphibian drifted idly on the surface of the water. There were green banks on either side of us, dotted with trees, some tilted at crazy angles, huts, and startled, craning people. Each of us in the amphibian felt languorous, inclined to lie still and relax, as though after a satisfactory act of love. Rehman lit a stubby pipe, pointed down at the river, and remarked, "It's full of sharks and crocodiles, you know. The sharks come up from the sea for the corpses."

A rowboat came towards us from the left bank. "That'll be the Minister's luggage," Rehman said. "We'll load it on, and the boat will ferry you back." The boat came alongside, sunk to the gunwales. It contained three large suitcases, two airbags, a holdall, an impressive briefcase, and a roll of bedding. "How long has this chap been here?" I asked. Rizvi smiled a cryptic smile. "Two days," he said, "but he's a Minister, you know."

The boat, divested of its burden, sped me back to shore. The Minister stood in a very ministerial pose on the deck of the cabin launch where he had apparently spent the night. He was attended by a valet, the local District Commissioner, and a number of police officers. I had been told to ask for the District Commissioner, so on clambering into to launch I approached him and told him my business. The Minister overheard me. He swung round, smiling affably, and extended his hand. "So you are a foreign correspondent? How fortunate for you that you have arrived at this moment. You may obtain from me a fine story of the details of my visit to Patuakhali. I have seen (you may quote this, please) the sufferings of the good people in this district, and I have shared them, both physically and spiritually. Please write all this down. It is well worth your while."

He continued in this strain for some minutes, then wrung my hand and climbed, apparently satisfied, into the rowboat. It paddled him slowly out to the amphibian, which he entered, pausing unsteadily on the step to wave a hand shoreward. The amphibian belched into life, threw up an angry wake of water, and flew off to Dacca. The District Commissioner watched its departure without comment. There was an ambiguous look in his eye.

Cattle and people disputed the narrow, dusty alleys of the little town. Beyond it, up an unmade road, stood a squat concrete structure which housed the Commissioner's office. He sat down at his desk, a dapper young man, his eyes shielded by tinted spectacles, and offered me an American cigarette. Then he lit it for me, took off his spectacles, and blinked his large liquid eyes, which unlike the Minister's were rather tired, several times in rapid succession. "Patuakhali district," he said, "has an area of 1613 square miles. There are — I should say, there were — about 150,000 people living in this area. According to the official statistics, 42,000 people were killed in the disaster, as of today. As of today, 866 are missing. There are enormous numbers whose deaths have not been reported — fishermen and people in small craft, the people from Sylhet who come once a year to cultivate the offshore islands — and bodies are still being found. So I would say one could safely estimate bet-

been 50 and 60 thousand dead. That is to say, more than one-third of the population of Patuakhali has been killed."

He was something of a poet. He lifted his long fingers, a cigarette smouldering down to white ash between them. "Who has counted the dead? Some, the earth has swallowed them up, some, they are floating in the sea, some, they are in the bellies of the sharks. Their trouble is over, but for those who remain the real trouble has only begun now. They have lost everything except their lives. Many of them may feel that they are better dead. But basically they are courageous people. When a little of the shock has died, I think they will fight for life once more. The Government will subsidise them with small amounts of money, new seeds and fertiliser, and materials to build their houses. It is the mental shock they have received which will be the most difficult thing to fight."

At this point a fat, elderly man in khaki entered the office. He was introduced to me as an important police officer of the district. "No," he said in answer to my question. "The police have not had much to do, so far as the disaster is concerned. In the first days before outside help arrived, my men helped with the distribution of food. But now there is nothing for us to do. We are not doctors or nurses, our purpose is simply to keep people in order, and now there is no necessity for us to work, because nobody is disorderly." "For the police force," he added, warming to his topic, "the cyclone has been very lucky. This is normally the time of the year, as the harvest is reaped, when most crimes take place. Murders are mostly committed over crops or land, and of course women too, because this is winter and the people are more active. But this year the people are too preoccupied with how to survive to think of murdering others. Yes, I tell you, this cyclone has been very lucky for the police, and we are all on holiday." He rubbed his hands together, with a noise like two scraped biscuits, and said, "It is the Commissioner who has to worry."

The Commissioner smiled, a wry twist of the lips. He started to describe his own difficulties at the time of the cyclone. "This," he said, "is the largest town of Patuakhali district. 18,000 people live here. I am not of this place myself, I come from Dacca. The town was too far inland for the cyclone to hit it with much impact. Still, it hit quite hard, though only fourteen people were killed here. But the morning after the cyclone I knew that the havoc down the river must be terrible. I could not reach this office, because of the floods and the wind, until the afternoon. A few other officers also managed to reach the office. We found that we could establish no radio or telephone communication with anyone anywhere. The town is not connected by road to any other place, and we had no boats, either to send to Dacca for help or to proceed to the costal areas and the offshore islands to see what had happened. All we could do was sit and wait. We sat in this very office, without food or drink, for 24 hours. Then an Army helicopter came. I flew in it over the coast and the islands. I could not believe my eyes. So many dead. . . . They were floating in the water like drowned flies. And where the villages had been, there was nothing. I had not dreamt it would be like that. Bad, yes, I knew that it would be, but not a catastrophe of that kind. When I returned to town, I opened the emergency rice stores. I bought rice from every merchant in town to add to the Government supply. Next day there was some transport, and we were able to take rice to some of the nearer areas. Afterwards proper help started to arrive: doctors came, and the Army, and after that the British Marines."

"They have been wonderful here," the policeman said. "They have done wonderful work. That is something which the Western people, the British, the Americans, have: they have this kindness, they are ready to help. But our own people, they are not kind. They do not care." The Commissioner chimed in at this point. "He means, the British and Americans are sensitive, they have heart. They value life. You know, I read a lot, as much as I can in this place—I have a complete file of the *Reader's Digest* for last year—and from reading Western books I have found out how much Westerners value life. It seems to me that in Asia we are not sensitive in that way, we do not value life, because there are so many people who are born and die every day in Asia." He nodded, pleased with his assessment. I said, "But the Indians are helping too. They gave you money, and supplies, and they offered you helicopters which Yahya Khan refused." The Commissioner's thin, handsome face puckered a little, as though at an unpleasant smell. "That is different," he said slowly. The police officer slapped his hand on the tabletop. "The Indians help us for their own reasons," he told me. "If we had allowed their helicopters in, they would have set up a Fifth Column here. Who can trust a Hindu? My grandfather told me what the Hindu *zamindars* did here in Patuakhali. Perhaps we may trade with them if Sheikh Mujib wins the election, but that does not mean that we should accept charity from the Hindus."

The Marines and the Pakistani Army both had their headquarters in the old Circuit House in Patuakhali town. The garden was full of neat khaki tents, outside which sat a number of young Englishmen, lobsterish in the sun, and naked to the waist. They read paperbacks, smoked, and waited to be called to duty. Two of them sat on the Circuit House wall, making friends with an uncomprehending crowd of locals. In the control room the wireless sets were croaking out reports from the chopper pilots already out above the islands. Here I met Colonel Bailey. He was tall, loosely built, and handsome, dressed in fatigues to which he contributed a certain elegance, and he invited me up to his room for a beer. Over the beer, which was Malayan Tiger, he said, "We're stationed in Singapore. Our Brigadier came out here a short while after the cyclone for a recce, and by 24 November we'd moved in to Patuakhali. We brought our landing craft and choppers. The Brigadier went back to base a few days back, so now I'm in charge. Natural disasters are part of our duty, you know, in peacetime. I've been 25 years in the Service and I can recall being called out for four of them—earthquakes, floods, and so forth—in that space of time. So the men who had already been a good while in the Service were pretty well used to this kind of work. The younger lads buckled down to it nicely from the start. What do we do? Well, at first we helped the Pakistani Army bury the dead. Now we're mainly busy on food distribution, which we also did from the start, and on maintenance work, keeping the Pakistani boats in shape. We repair them all the time, but the curious thing is the Pakistanis don't seem to be very mechanically minded: you repair a craft and hand it back to them, and in five minutes the engine's dead again. Apart from that, the Pakistani Army is excellent to work with. They liaise between us and the officials. That is, if the officials want something done, the Army comes and asks us to do it, and we tell them if we can or can't. They've done a lot of fine work here."

Colonel Bailey's blue eyes looked thoughtfully down into his beer. His room was crowded but orderly; several pairs of boots under the bed, a photograph of his wife on the dresser, his kit on a hanger by the cupboard : the room of a serviceman on duty away from home. "Communications here are appalling anyway. Of course the cyclone didn't make them any better. I think that was the reason there were so many deaths, because nobody in Dacca knew about what had happened down here for days after. But I think the situation's under control now. We've already started to pull out, and we'll all be out by 12 December. We've 500 men here, and we have to move them bit by bit. I think from now on the Pakistanis will be able to handle the situation. They *ought* to be able to handle it."

The slow waves creamed around the islands, and as the chopper clattered overhead the people looked up, shielding their eyes against the sun. Their villages had been wiped out, but neat temporary structures of bamboo and thatch had been put up, and between the huts were spread cloths on which rice had been put out to dry. The shock was fading, perhaps, and the urge to life had started to utter its usual imperatives. My pilot, a slight, crewcut Cockney, looked down at the islands as we went over. "Poor buggers," he said. "'Ad a rough deal, ain't they? Lookin' round these parts, you can see why the Pakis want to emigrate to England, can't you, sir?" We descended to an island shaped like a muddy pear.

"Want to do y'stuff, sir?" inquired Ginger when we were down. He pointed at the people racing towards the chopper. "Talk to the chaps, like?" I said I'd have liked to, but couldn't, since I didn't speak the language. The villagers crowded round me as I climbed out into the sunlight. They were skimpy, dark people, and they seemed excited and pleased, not only because we brought supplies, but because we came from the outside world, which was manifesting an interest in them for the first time ever.

One of the men, while Ginger was manhandling the square boxes of food and blankets over to the other villagers, nodded to me in a friendly way. He spoke a kind of English. He waved his hand at a woman standing a little way off, her sari drawn over her face. "That one, her man, her children, all dead. Now nobody look after her. That man there, all fam'ly dead. My mother, father, sister dead. Mostly women, children, old people, dead. Can't swim, can't climb tree. Man can do. When water coming, man swim, man climb tree. Woman, children, old people, no can. So—phoosh!" He imitated a wave sweeping hundreds of the frail and helpless out to sea. It was rather like a man describing an exciting cricket match in which he himself has played. I was glad when we took off.

Every island was the same. Each had its terribly depleted population, its newly raised shelters built by the Pakistani Army, Ansar volunteers, and the Marines, and the people on each showed the same pathetic eagerness to welcome us. But they had their subtle differences: the only tree that survived on one of them held its damaged arms plaintively up towards the sky, and perched among them, looking utterly absurd, like some deranged and dilapidated angel, was a small boat whose ragged sails fluttered like wings. On another island a woman who stood apart from the other villagers was pointed out to me. She had not only lost her husband and children, but every single relative she possessed. With the position of women what it is in Islamic society, her future was obviously dismal in the extreme. She looked hopefully at the helicopter, Ginger turned his head away.

"It's funny, innit, sir," he said when we were up once more. "There's so many people in the world a bloke could do a little for and not feel it, and it would be a terrific chance for them, and a bloke never does it somehow." He removed a cigarette from behind his ear and lit it. We were flying low across the corner of an island, and in a shallow pond below I noticed two long unidentifiable shapes, quarrelled over by sleek crows. "Corpses," Ginger said laconically. "There were plenty of those around when we first arrived." He wreathed himself in smoke. "Cor," he said. "It didn't 'alf pong."

We returned to the town. The Commissioner was waiting for me in his office. It was afternoon, and a low hot sun burnt sullenly in the western quarter of the sky. "The only way for you to return to Dacca," he said, "is by the river steamer. It is about to leave but I have reserved a seat for you. It will take, I am afraid, sixteen hours, but there is no other way." His jeep bumped me hastily down to the river, where, unsteady on its hawsers, an extremely battered steamer, coated in rust, awaited departure. It had two passenger decks, which were packed solid with peasants, sitting or lying on the floor: a passage through them had to be accomplished by main force. On the upper deck were two small cabins, each of which contained about ten people wedged together on wooden seats. I was wedged by the policeman the Commissioner had sent with me into one of these cabins. The boat, which I had noticed was called "High Speed," snorted, cleared its throat, and began to shuffle very slowly upriver towards Dacca.

The landscape which we passed was one of the most beautiful, I think, I have ever seen. It was very green and very flat, but trees nodded slowly over the river as it coiled idly by, dropping their long shadows over the turf and the crops. The few people on the banks shouted and waved cheerfully as the steamer passed. At evening, in the scented dusk, one could see their small fires on the shore, and the tall tilted silhouettes of the trees that followed us up the river. It was a pastoral landscape, and a pastoral way of life, that had endured for centuries, and it seemed absurd that a countryside that so obviously mirrored the flow of human life should have been hit by so inhuman and uncontrollable a force as a cyclone.

At Barisol, the port of Bhola, eight of the ten people in the cabin left. The remaining passenger beamed at me, shook my hand, told me that he was an Army doctor from Patuakhali, and inquired who I was. When he had learnt all he could possibly learn about me he shook my hand once more. "So you are British," he said, "but you come here from India. How do they treat the Muslims in that dirty Hindu country?" I said I thought the Muslims were treated reasonably well. "What about Ahmedabad?" he retorted. "Last year those dirty Hindus killed five million Muslims there." I told him that this was not true, that I had been to Ahmedabad myself, and that though the whole incident had been deplorable, not more than two thousand Muslims at most had been killed. "Of course, the Hindus would not tell you the truth," he said. "But I heard it on the radio. Let us not talk of these dirty Hindus. You are my guest. You have come from a long way to my country. Now the accomodations on this boat are very bad. There is no food here, *and* no water, *and* no bedding. But *I* have all three. You will share my food and water, and I will lend you a pillow and a blanket." He overruled my feeble and rather false protests, and we dined off fish curry, spiced potatoes and rice. The doctor then opened the cabin door, to visit the lavatory. This was something I had resolutely decided not to do, since one lavatory served the whole ship. Outside the door lay a young, and perhaps once beautiful, woman, clutching a very small and fragile baby to her breast like a child with a doll. Her eyes were shut, and she was moaning spasmodically. The doctor bent over and looked at her for a while. Then he straightened up quickly, and stirred her with his toe. He addressed her angrily in Bengali. Very slowly, without a word, she rose, clutching her child, and moved towards the passenger deck.

When the doctor returned, beaming once more, I asked him who the woman had been. "I do not know," he said. "But I think she was suffering from cholera. It was not healthful for us that she should sleep outside the door." I protested that if she really was suffering from cholera, it was up to us to do something. "No, no," said the doctor, "in this large ship we will not be able to find her anyway. Besides, though I have the greatest sympathy for the cyclone victims, the peasants live like dogs, and they die like dogs. Her case is nothing unusual." He occupied the rest of the evening, until we went to bed, in showing me photographs of his wife and four small children.

Chapter 4

CLOUDS ABOVE DACCA

A heavy hand on my shoulder roused me from an unquiet sleep. I peered blearily up into the broad placid face of the doctor. "We have reached," he said. "But we are very early." A glance through the porthole showed me what he meant. It was pitch dark outside, except where the uneven flare of kerosene lamps revealed the interior of a quayside shed. A sepulchral mist shrouded the river. I shook the doctor's hand, said goodbye, and plunged through the cabin door. The narrow corrior outside was packed solid with people. They were fighting their way towards land with an intensity which suggested that they feared that the ship was about to sink. Children wailed, adults shrieked, and a powerful rancid smell of bodies filled the air. I battled my way slowly through to the ramp that led down to the quay.

The alleys beyond the harbour exit were providentially cool and still. Two or three cookshops, illumined by kerosene lamps, exuded steam and odours of food. The passengers off the steamer poured avidly into them. I went on to where, in the throat of the alley, a solitary cycle rickshaw stood, climbed into it, and told the driver where to take me. We set off, in a slow and meandering style, through a sequence of narrow, unlit, totally deserted streets. I started to feel nervous. The driver, however, showed no tendencies towards murder or theft: he wheezed asthmatically over the pedals, and I eventually felt rather sorry for him.

We came out of the old city at last, and proceeded at a smart clip down the wide boulevards towards the massive bulk of the Intercontinental, lit like an ocean liner on the dark sky. The streets were deserted, even here, except for a few quiet defecators in the shadows. I remembered that today Pakistan went to the polls. Nobody seemed very excited about it, this early in the day. The chalked symbols and occasional posters on the walls looked spectral under the horned moon. At the Intercontinental, I climbed out of the rickshaw, paid, and limped upstairs, hoping to be able to sleep, bathe, and have breakfast in bed.

My hopes were shattered when Wahid telephoned at seven. "This is all very exciting," he said. "You must see the people at the polls. It will give you a good story. Perhaps there will be riots, murders, arson. I do not think so, but who knows?" I dragged myself out of bed, had a rapid and unsatisfactory bath, dressed, and limped back downstairs to find Wahid awaiting me. He insisted that we should return to his flat for breakfast. A number of the other young men were already there.

We drank tea flavoured with cardamoms, and ate sticky

sweets. This was not exactly what I had in mind for breakfast. As we ate, however, everyone suddenly ceased to speak. They all looked delightedly at the kitchen door, and expectantly at me. When I looked towards the kitchen, a young woman stood in the doorway, clutching a baby to her hip with one hand and a small boy to her side with the other. An *odhni* covered her face. "My wife," said Wahid with considerable pride. "She made these sweets." Before I could think of a word to say, the young woman slipped back into the kitchen. "Now," said Rahman sententiously, "Wahid has proved that he is your friend."

We sat on, talking. Nobody made any move to leave. This was not, my conscience told me, the way a foreign correspondent should behave. Ideally, at this moment, I should be racing round the polling stations trying to spot trouble. But I had no desire to do so. I listened to the young men instead. They talked of their relatives abroad—most of them had some remote kindred in England or America—and of how they would like to visit a foreign country. "China, I suppose," I teased, but they did not smile. "No, a rich country, England, America," Wahid said. "We could make money there. Perhaps that would be bad. If we made very much money, perhaps we would not wish to return to Pakistan." But they continued to speculate on which country would offer the most opportunity. "Because," one said, "if we remain in Pakistan, we will remain poor. Even though we are Government servants." Everyone laughed. "*Because* we are Government servants," Rahman said. "The Government does not pay properly to anyone. Look at the peasants in Bhola and Patuakhali. They should pray to Allah to send a cyclone like this every year. It is the first time the Government has ever fed and clothed them free, or taken any interest in them at all."

Wahid said, "This is the only political cyclone in history. Look how it has changed our politics. We are a Muslim country but we should adopt the Christian calendar. BC and AC. efore the cyclone, and after." Downstairs a radio belched out news bulletins, and the noise in the street slowly built up. I suggested that we should visit the polling stations. "Yes," Wahid said. "We must remember our duty as citizens and cast our votes." I asked if his wife intended to vote. "No, no," he said. "How could she know whom to vote for?" We climbed down the resonant stairs into the full day. By this time, though the shops were all shut, steel eyelids sealing off the windows, the roads were full of people, on foot or in cycle rickshaws. The pedestrians floated along in talkative clumps. They were all headed the same way, but had a rather aimless look about them. Those in the cycle rickshaws were often *burkhaed* women, with children in their arms. "Will they vote?" "Yes," said Wahid, "their husbands will have told them who to vote for. But I, I have too much respect for the freedom of my wife to dictate to her like that."

The situation exemplified part of the problem of Pakistan. The rich and educated Pakistanis, as a rule, unwind their wives from the *burkha,* make them paint their faces, polish their toenails and their manners, and attend cocktail parties. But, come the elections, the wives probably still vote for the candidates their husbands approve of. This is a liberation of the body rather than the mind. For an intelligent but poor man, the difficulty is different. He may want to liberate the body of his wife, but the society in which he lives forbids this. He may want to liberate her mind but first he has to create a mind to liberate : he has to try and educate her. This becomes doubly difficult because he is usually still trying to educate himself. Hence the dichotomy between the liberal mind of someone like Wahid and his domestic circumstances.

One polling station is much like another, all over the world. The three that I visited had no unusual features about them. The queues outside were orderly. They squatted in line, looking thoughtful, and sometimes munched food out of leaf containers. The officials I talked to did not expect any trouble. "Why should there be trouble?" one of them demanded. "If there was any opposition to Sheikh Mujib, then perhaps we should expect it, but there is no opposition really. So it will be very peaceful today." The sun was very hot, and I had started to feel very tired. So when Wahid and his friends went off to vote I returned to the hotel. The huge lobby was empty for once, except for a local newspaperman sipping beer. He informed me that all my colleagues were out, keeping an eye on developments. I said it was probably very unprofessional of me not to be doing exactly that. "Why?" he asked, his eyebrows quizzically lifted. "There are no developments to keep an eye on." Comforted, I went up to my room, switched on the airconditioner, and fell asleep.

It was afternoon when I awoke. I hired a taxi and went for a ride round the city. The heat of the day had faded, and there were more crowds than there had been earlier, hurrying on foot or by cycle rickshaw to cast their votes before the polling stations closed. On the pavements, groups of those who had already voted listened to forecasts and early results on transistors. Radios blared in restaurants, which were packed like concert halls with attentive auditors. As the cool, scented dusk came down, and the lights of the city flicked to life, I went back to the hotel. Most of the other correspondents, hot and tired, were in the bar. A deeply depressed American, who had heard the previous day that there would be a bloody riot, and had spent precious and now wasted hours polishing a pile of phrases to describe it, kept asking people if they had seen any trouble at the polling stations. A Japanese photographer said yes, he had. The American came abruptly to life. His hand flew to his notebook. "What did you see?" he demanded. The Japanese replied, "I saw a man arrested because he tried to vote twice." Neither he nor the American understood why everybody laughed so much about it.

Throughout the next day the results continued to be declared. The East Pakistanis, riveted to their radios soon discovered that Mujib had won even more seats than anyone had expected. Nine seats in the cyclone area would be decided later, so only 153 were contested. Of these the Awamis captured no fewer than 151. The nine flood seats were pledged to Mujib, and with the seven nominated seats for women, the likelihood was that he would hold 167 seats in the National Assembly. A month after the elections, the death of one of the two non-Awami members caused a bye-election which was won by the Awamis. Mujib thus stood to hold 168 seats in all out of 313 in the Assembly.

Bhutto, in the West, had been considerably less successful. Of the 138 Western seats, the People's Party captured 82. That is to say, it held less than half the number of seats held by the Awamis. Mujib's position, in any other country, would now seem unassailable, but the fluxes and shifts in Pakistani politics have been so unpredictable in the past that one hesitates to foretell what will happen in the future. Also to be considered is the fact that Yahya Khan will have to approve the new Constitution when it is finally drawn up.

Mujibur Rehman has defined his stance clearly. At the

Round Table Conference of March 1969, he said: "If the demands made by different sections of the people are carefully examined, it will be seen that there are three basic issues that underlie them. The first is that of deprivation of political rights and civil liberties. The second is the economic injustice suffered by a vast majority of the people, comprising workers, peasants, low and middle income groups, who have had to bear the burden of the costs of development . . . while the benefits of such development are increasingly concentrated in the hands of a few families, which in turn are concentrated in one region. The third is the sense of injustice felt by the people of East Pakistan who find that under the existing constitutional arrangements their basic interests have consistently suffered in the absence of effective political power being conferred upon them. . . ." These three issues are the ones he seeks to solve. They have directly to do with the future of Pakistan.

Bhutto's stance is more ambivalent. He was once Minister of Commerce under Ayub, but he violently opposed the Tashkent Treaty of 1965, and crossed over to the opposition in that year. In 1967 he founded the Pakistan People's Party, which according to its manifesto stood for the restoration of democracy and adult franchise, and which declared itself socialist but "with Islam as our basis," which is a contradiction in terms, since Islam holds private property to be sacred. He was imprisoned for disrupting the peace in 1968, but released the following year. Bhutto has so far appeared sympathetic to East Pakistani aspirations. During a visit to Dacca in 1966, he described the eastern wing as "the most important part of Pakistan." Two years later, in London, he declared that the administration under Ayub had created a vast economic division between East and West. He supported the East Pakistani demand for self-government. It is difficult to ascertain how far Bhutto is serious, and how far he is a weather-cock for political winds. Certainly he is very consistent in his hard line towards India. In the course of the 1965 war, he said that Pakistan "would fight India for a thousand years for the liberation of the enslaved Kashmiris." Not all his quarrels with India are on so large an issue : his most recent exploit has been to threaten to dig up the hockey field in Lahore if the Indian team was permitted to play in the World Cup Tournament there.

From the Indian point of view, it would be an immense help to have Mujib at the helm of Pakistan. He is an intellectual, liberal in all his views, and, which is more than one can say of Bhutto, consistent in his policies. What he wanted in 1969 is what he still wants. He wants the establishment of a federal parliamentary democracy, with the representatives of the people chosen by direct election and full adult franchise. The representation in the federal legislature is to be on the basis of population. "The principle of parity in each wing," he said in 1969, "...is based on the false premise that representatives ... are likely to vote on a regional basis.... East Pakistan has always subordinated its regional interest to the overriding national interest."

He took a firm stand, in 1969, to obtain what he wanted. He pointed out that East Pakistan had consented that the federal capital and all the defence headquarters should be located in the West. This meant that the bulk of expenditure on defence and civil administration — 70 per cent of the Central Budget — was incurred in West Pakistan. If the West Pakistanis refused the demand for legislative representation on a population basis, Mujib said, the East Pakistanis would be compelled to demand that the capital and defence headquarters be shifted to the East. He produced, during the same speech statistics to demonstrate the disparity between East and West.

"As early as 1960," he said "...the real per capita income disparity between East and West was 60 per cent ... it has been steadily increasing." (In 1959-60, in fact, the per capita income in the East was Rs. 278, in the West Rs. 366. By 1966-7, the per capita income in the East was Rs. 313, in the West Rs. 463.) The capacity for power production in the West, he added, was "5 or 6 times higher than in East Pakistan, the number of hospital beds in 1966 in West Pakistan was estimated to be 26,200, while that in East Pakistan was estimated to be 6,900 during 1961-66 : only 18 polytechnic institutes were established in East Pakistan as against 48 in West Pakistan.... More than 80 per cent of all foreign aid has been utilised in West Pakistan in addition to the transfer of East Pakistan's foreign exchange earnings to West Pakistan. This made it possible for West Pakistan over 20 years to import Rs 3,109 crores worth of goods, against total export earnings of Rs. 1,337 crores ... during the same period East Pakistan imported Rs 1,210 crores worth of goods against total export earnings of Rs. 1,650 crores." To remedy this appalling situation Mujib produced a four-point plan.

He suggested first that two separate currencies should be established, or failing this, that a separate Reserve Bank to control monetary policy should be set up in each wing. This would prevent "the flight of capital from one region to another." Secondly he proposed that each regional Government should be able separately to negotiate trade and aid within the framework of the foreign policy of the country, which would be the responsibility of the federal Ministry of Foreign affairs. Thirdly, he said, the foreign exchange earnings of each region should be held by the

Reserve Bank of that region, and be controlled by the regional Government. The federal requirement of foreign exchange would be met by appropriations on the basis of an agreed ratio from each regional account. Finally, taxation would be the responsibility of the regional Government, but the federal Government would be empowered to realize its revenue requirements by levies on the regional Governments.

These proposals are reasonable and sensible, and while reason and sense have not played a very significant role in the past politics of Pakistan, the country has now arrived at a point where the two wings, to quote Auden, "must love one another or die." The East cannot be put upon any longer, nor can the West afford the only other solution : if it does not accede to the Eastern demands, Bengal may eventually try and break away into an entity outside Pakistan. It is essential to the future of the country that it should not do so. Pakistan, as a political power, depends on the fact that it is the most populous Muslim country in the world. Out of an estimated 125 million people, 70 million live in the East. If the two wings split, neither would in itself be sufficiently large to possess the international pull of a united Pakistan. The second factor is that while West Pakistan forms part of the Middle East, East Pakistan forms part of South-East Asia. Geographically therefore, Pakistan as a whole has interests in two spheres. Moreover, East Pakistan earns foreign exchange for the country as a whole, and provides a market for the industrial products of the West. Given a situation which has not so far existed, where East and West are equal and interdependent, there is no reason why the country, despite its unique physical division, should not function as an efficient and peaceful whole.

The reasons why it has not hitherto done so have been twofold. The economic drain on East Pakistan has ensured that the wealth of the country is in the West. The Government policy, dictated from the West, pushed more and more perquisites into the hands of the notorious "22 families," of whom the London Economist wrote, "They have grown often with little enough to start with, after fleeing from India, through cotton and jute into rayon, chemicals, paper, fertilisers. . . ." The Government helped them, particularly under Ayub, whose son Gohar was financially involved with some of them, through a differential tax in favour of paying dividends, a rebate on personnel income tax for investments, and a six year tax holiday. The reason brought forward for this by the Finance Minister was that it promoted capital formation, and filled the lacuna left by the migration of Hindu and Sikh industrialists from West Pakistan. But basically the interests of the 22 families are those of the Army : so long as the people are kept under, there is profit to be made. The East, however, was left to itself, and the growth of industries there was not promoted. The excuse for this was that all the private capital was in the West, and Western industrialists were not anxious to invest money in an unexplored field a thousand miles away. But this economic unfairness to the East had its roots in the second reason for the lack of cohesion between the wings of the country.

This reason is the attitude of the West Pakistanis towards the Bengalis. The Muslims of the west are intensely conscious of their Islamic heredity, and the establishment there is mainly descended from the bloodthirsty Afghan and Arab invaders of North India. The East Pakistanis, by contrast, are the descendants of Hindu converts. The language they speak, Bengali, is a Hindu language derived from Sanskrit, and their culture is Bengali too. The Westerners speak Urdu, which is a Muslim language derived from Arabic, and their culture is Middle Eastern. They have therefore, at least in their own opinion, a spiritual edge over the Muslims of Bengal.

The curious and barbarous habit, widespread in India and Pakistan as well as in South Africa and the less civilized parts of the United States, of considering a dark man inferior to a fair one, operates in their minds. The small, dusky Muslims of Bengal seem to the West Pakistanis physically and spiritually inferior. In the days before Partition, the two communities hardly ever came into contact, except politically, and there the Eastern leaders were aware that the Muslims of the West were not really interested in the welfare of Muslim Bengal. Since 1947, the mass of the East Pakistani population seems to have realized that the Westerners consider them inferior. They have, in fact, suffered the effects of this economically, in much the same way as the blacks in South Africa or Alabama. They have earned money abroad which has been taken away to benefit the West: they have been denied an opportunity to improve their situation. The point of rebellion has been reached, and it is for the West to recognize this.

If the demands of the East are met, the economic disparity between the two wings will be remedied and the superiority complex of the West Pakistanis will cease to constitute an irritant in an autonomous state. If they are not met, anything may happen. The ancient, bearded presence of Maulana Bhashani appears to have been forgotten by many commentators. In 1963, Bhashani headed an official delegation despatched to China by Ayub. The achievements of China so impressed him then, he said, that he realized how backward Pakistan was. This inclined him, he announced at the time, to "spend the rest of his life in prayer." Unfortunately for everyone, he failed to do so. Instead, he turned his NAP into a pro-Peking party.

The rural areas of East Bengal are ripe for something like the West Bengal Naxalite movement, and indeed a certain limited amount of Naxalite activity has been reported there. In 1966-7 the per capita rice output touched a new low, and in 1969 the price level of rice rose by 30 per cent. 75 per cent of the East Pakistani population is employed on the land, and 36 per cent of the peasantry is made up of landless labours. 65 per cent of the landed peasantry, who own less than two acres apiece, is in much the same boat as the landless labourers. The Director of the East Pakistani Academy for Rural Development characterizes the East Pakistani peasant as "suspicious, full of distrust and fear." Rehman Sobhan, a leading East Pakistani economist, says that the Bengali ryot "awaits the ideas and organisation which can make him aware of his latent political power. If this catalyst is ever injected into the rural areas, one can expect a social convulsion . . . which could engulf the regime itself. . . ." The NAP, in 1967, set up an East Pakistan Peasant Society with branches all over the rural areas. This may have been designed to provide the catalyst which Sobhan fears.

The potential power of the NAP is increased because the East Pakistani middle class, adversely affected by what it thinks of as the colonial policy of the West, is not alienated from the peasantry. On the contrary, both classes share a dislike of West Pakistan which the cyclone havoc has intensified. They (rightly) blame the Centre for the absence of flood control and the failure of communications which delayed the advent of aid. The combination of a disaffected middle class and a peasantry filled, as a Pakistani observer has put it, with "frustration, bitterness, and distrust," is the classic formula for a popular revolution. Sheikh Mujib is the only Pakistani politician who can fight this fairly, that is to say without the use of force. His programme to nationalize industry, banks, and insurance companies, the owners of which are mostly West Pakistanis, has immediate popular appeal. He has not repudiated the role of private enterprise, so that this would mean elimination of West Pakistani competition in the East. His policies, if pursued, would logically lead to the enrichment of the middle class, and thus knock one of the principal props out from under the NAP. His charisma, at present, works mainly in urban areas, but if he were able to implement his policies, would probably spread down through the middle class to the peasantry, now solidly behind Bhashani. The Maulana knows this. He recently changed his tune about autonomy, and demanded instead a fully independent East Pakistan. This threat of secession may well serve as a pretext for the Army to deny Mujib his reforms. They would affect the military financial autonomy, and neither Yahya Khan nor the 22 families wishes to see them become part of the law of the land. In this, at least, Bhashani concurs with them.

Ironically, Mujib's clean sweep in December was partly due to the fact that the NAP did not contest the elections. It was the chief rival to Awami leadership, and had been expected to capture a considerable number of seats both in the National and Provincial Assemblies. The hardline Marxists in the party, however, felt that participation in the elections would constitute a tacit acceptance of bourgeois democracy. They split with Bhashani over the issue. He decided, therefore, not to contest the elections. He had a splendid and readymade excuse to hand : he implied that the collective heart of the NAP bled so copiously for the cyclone victims that its members had decided to dedicate themselves to relief work rather than elections. In the absence of the NAP a number of its supporters voted for the only other party to demand autonomy for the East. Whatever the causes that motivated the electorate, however, Mujib now has a mandate from the people of his province. If he is denied the opportunity to implement it, the most likely consequence will be a civil war between the East and the West, culminating in the military occupation of the East. If he is given the opportunity and fails, the most likely consequence will be a Communist revolution. Either outcome will mean the final collapse of any hopes that Pakistan can finally, after twenty years of strife, achieve a national identity.

The only real proposition that can be made in support of a religious state is that the concept behind it is a very antique one. This is also the best argument against it. What worked a thousand years ago will not necessarily work today, even amidst a primitive and backward people. The genesis of Pakistan, however, was unfortunate, in that the country was born out of religious differences. It was therefore, in a way, appropriate that it should declare itself a religious state. Islam was the only factor common to East and West alike. The founders of Pakistan thought it would be a binding factor. It was not. The East Pakistanis saw the West Pakistanis as foreigners who happened to share their religion, and vice versa. The Koran could not overcome the differences of language and culture, and above all the differences of interest, which divided the two people.

I have talked to a number of Indians who hold the opinion that the only common factor which unites the people of their country is the Hindu religion. This displays a certain arrogance of attitude towards the followers of other faiths who also happen to live in India. Moreover, it does not square with another assertion made by the same people, which is that the country is an immensely tolerant one. But then consistency has never been a hallmark of the Indian character. The only powerful effect religion has on people today is to divide them, not to unite them. A religion in this century, like a colour of skin, can mark a man out for attack. When I watched the 1967 riots in Detroit, the two sides were of different colour, so that each could clearly identify the other. I was present, three years later, at the riots in Belfast: there the Protestants knew who the Catholics were because they lived in certain specific areas. Colour in one case, and religion in the other, marked out the enemy, but in both cases the participants were underprivileged people, full of undirected hostility. The Belfast Protestants, for example, were not particularly fervent in their faith, but a tribal instinct made them resent the nearby presence of an identifiably different tribe, and frustration and poverty did the rest. The black and white blocs in Detroit were internally divided, as were the Protestant and Catholic blocs of Belfast. Each bloc was unified only by the identifiable difference of its opponents from its members. Similarly the Pakistanis are not unified by Islam, but hate India because they know the population is mostly Hindu, a different tribe and therefore undesirable. Religious and racial superiority are both dead concepts; both have lost their original purpose. Once they were props for the structure of a hidebound society; now they are rotted sticks ripped out of the structure to beat other people with.

Those who still feel that a religion can unite a country would do well to consider the case of Pakistan. Its founders thought that a poor and backward people would rally to the call of religion more quickly than to the call of nationalism. The Islamic Republic echoes with the name of Allah: he is invoked daily from minarets and political platforms: but he has not so far cared to reply. The East and the West are still a thousand spiritual, as well as physical, miles apart. Nowhere can the concept of Islamic brotherhood have been disproved so completely as in Pakistan.

There is another parallel between India and Pakistan. East and West Pakistan are physically far apart, the language and customs of their inhabitants are completely different. The differences have overriden the common link of Islam, and led to serious trouble. The linguistic riots of 1952, when the Centre attempted to make the Bengalis speak Urdu, are an example. But, as a Dacca businessman remarked, Amritsar and Madras are as far apart as East and West Pakistan, their languages and customs are completely different, and when after 1962 the Centre proposed that the Madrassis speak Hindi, linguistic riots were the result, as they would be were it proposed that the Amritsar Sikhs should learn Tamil.

Linguistic and regional differences have been the bane of Indian politics, but these issues, in India have been very complex, and diffused over a wide area. In Pakistan, because of the actual physical separation of the two wings, the same issues are relatively clearcut, and their solution is clearcut too. East is East, and West is West, and never, at least geographically speaking, the twain shall meet. Nevertheless, if the proposals made by Mujib are accepted and implemented, the twain will at least combine. Two autonomous provinces will work in a business partnership for mutual profit. Patriotism, on a national basis, will not really come into it: the patriotism involved will be provincial. But patriotism on a national scale — nationalism — is, like religion, an exploded concept. What is important is that a country should work, on the impersonal level that a poem or a machine works, and on paper at least, Mujib's programme seems to do so : if it doesn't, nothing else will in Pakistan.

The concept of nationalism is important to the new Asian and African countries because in many of them it has been a comparatively recent development. But the concept is outworn because, like religion, it leads to a certain narrowness of outlook, a chauvinism, a revulsion from unacceptable reality. Nationalism in its worst form too often turns a country inward on itself, cuts it off from the rest of the world, and blinds it to its own defects. It is an essential force in any freedom movement, but a disaster in an evolved and independent state. Some of the newly evolved states of Asia and Africa, still close to a time when they had to fight for freedom, have not yet realized this.

If the proposals of Sheikh Mujib form part of the new Pakistani Constitution and are transmuted into law, they will constitute a denial of the concept of total national solidarity. But this will be realistic, and will provide the basis on which a viable country could rise from the ruins of Jinnah's original ideal. It is despair, ironically, which has driven Pakistan to face reality.

India, however, is in a position to exercise intelligent choice and free will upon her future. The two wings of Pakistan were once components of an undivided India. Brought into direct confrontation with each other after 1947, they clashed bitterly. There is no reason to suppose that the natives of Amritsar and Madras, drawn into a similar confrontation, would not clash likewise. The demands made for linguistic and regional separatism by local leaders in India since 1947 are anomalous in the context of an undivided nation. The concept of the undivided nation dates back to the freedom movement, but the demands, often backed by mass demonstrations, are still being made. The concept prevents a realistic assessment of the demands, at least by many of the more obscurantist Indian politicians. The concessions made to the separatists by Nehru and his successors have been intensively criticized by these people. Once the shadows of the nationalist dream are swept away, a created clarity may reveal that a federation of autonomous states under a weakened centre may be a solution for the problems of India as well as those of Pakistan. This would mean, as in Pakistan, an abandonment of the nationalist concept, but would not be a submission to chaos, rather a concession to necessity.

Once the elections are over, the press started to drain slowly out of the Intercontinental, as out of some huge marble aquarium. Those who stayed on gulped like stranded goldfish in the bar. I decided to leave with the rest. Since I wanted to buy some presents, Rahman, Wahid, and the others took me on a last trip round Dacca. They haggled fiercely with the shopkeepers on my behalf, but they seemed intensely aware that this was the end of my visit. As though anxious to put every minute to use, they bombarded me with questions about foreign cities—London, New York. Paris, what did they look like, what was the weather like, and finally, sadly, on a dying fall, how much did it cost to travel to these places. I suddenly realized that for them I was a bridge to these exotic capitals. They were there, vicariously, through me, perhaps the only way they ever would be there, perhaps for them the best way, for they would almost certainly be poor and unhappy in any foreign city they actually reached, like the Pakistanis who sweep the floors at London Airport. Their youth and frustration moved me. It must be terrible to be young and not to be able to realize any of your ambitions. It was no wonder they were Communists. I would have liked to have helped them, but there was really no way I could. "Goodbye," I said at the end, and climbed into the taxi for the airport. They stood around in the hotel porch. They looked very young and forlorn. Then the taxi driver, who was elderly and copiously bearded, leant over and spoke to them, and they burst into sudden boyish laughter. Rahman cried, "He says as Muslims we should not have entered your hotel, because here they serve liquor and pork. Ah, Mr Moraes, I tell you, this is Pakistan!"

The driver sulked in his beard all the way to the airport. An hour later I climbed aboard the Thai plane that would take me back to Delhi. We took off, flying high over the countryside I had traversed by amphibian and chopper and boat, an emerald and yellow sprawl of fields and water under the afternoon sun. What would happen to this curious, isolated land in the near future was problematic: and at the precise moment that the loudspeaker told us that we had crossed the border into India, a shroud of clouds, timed and symbolic as in the finale of a film, floated up under us, and hid East Pakistan from our eyes.

INDEX